O'NEILL

Forcible Entry

Seventh Edition

VALIDATED BY
**THE INTERNATIONAL FIRE SERVICE
TRAINING ASSOCIATION**

PUBLISHED BY
**FIRE PROTECTION PUBLICATIONS
OKLAHOMA STATE UNIVERSITY**

COVER PHOTO BY: BOB ROSE
Chico, California Fire Department

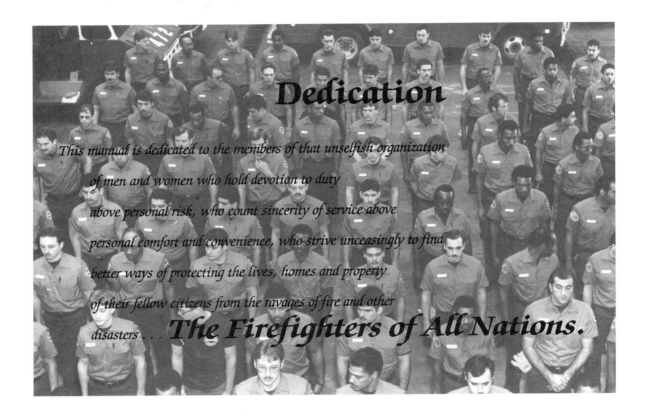

Dedication

This manual is dedicated to the members of that unselfish organization of men and women who hold devotion to duty above personal risk, who count sincerity of service above personal comfort and convenience, who strive unceasingly to find better ways of protecting the lives, homes and property of their fellow citizens from the ravages of fire and other disasters. . . . **The Firefighters of All Nations.**

Dear Firefighter:

The International Fire Service Training Association (IFSTA) is an organization that exists for the purpose of serving firefighters' training needs. IFSTA is a member of the Joint Council of National Fire Organizations. Fire Protection Publications is the publisher of IFSTA materials. Fire Protection Publications staff members participate in the National Fire Protection Association and the International Society of Fire Service Instructors.

If you need additional information concerning these organizations or assistance with manual orders, contact:

For assistance with training materials, recommended material for inclusion in a manual, or questions on manual content, contact:

Customer Services
Fire Protection Publications
Oklahoma State University
Stillwater, OK 74078-0118
(800) 654-4055
(405) 624-5723 in Alaska and Oklahoma

Technical Services
Fire Protection Publications
Oklahoma State University
Stillwater, OK 74078-0118
(405) 624-6832

First Printing, February 1987
Second Printing, March 1989

ISBN 0-87939-069-7
Library of Congress 86-82902
Seventh Edition
Printed in the United States of America

Table of Contents

List of Tables

THE INTERNATIONAL FIRE SERVICE TRAINING ASSOCIATION

The International Fire Service Training Association is an educational alliance organized to develop training material for the fire service. The annual meeting of its membership consists of a workshop conference which has several objectives —

> ... to develop training material for publication
> ... to validate training material for publication
> ... to check proposed rough drafts for errors
> ... to add new techniques and developments
> ... to delete obsolete and outmoded methods
> ... to upgrade the fire service through training

This training association was formed in November 1934, when the Western Actuarial Bureau sponsored a conference in Kansas City, Missouri, to determine how all agencies that were interested in publishing fire service training material could coordinate their efforts. Four states were represented at this conference and it was decided that, since the representatives from Oklahoma had done some pioneering in fire training manual development, other interested states should join forces with them. This merger made it possible to develop nationally recognized training material which was broader in scope than material published by an individual state agency. This merger further made possible a reduction in publication costs, since it enabled each state to benefit from the economy of relatively large printing orders. These savings would not be possible if each individual state developed and published its own training material.

From the original four states, the adoption list has grown to forty-four American States; six Canadian Provinces; the British Territory of Bermuda; the Australian State of Queensland; the International Civil Aviation Organization Training Centre in Beirut, Lebanon; the Department of National Defence of Canada; the Department of the Army of the United States; the Department of the Navy of the United States; the United States Air Force; the United States Bureau of Indian Affairs; The United States General Services Administration; and the National Aeronautics and Space Administration (NASA). Representatives from the various adopting agencies serve as a voluntary group of individuals who govern policies, recommend procedures, and validate material before it is published. Most of the representatives are members of other international fire protection organizations and this meeting brings together individuals from several related and allied fields, such as:

> ... key fire department executives and drillmasters,
> ... educators from colleges and universities,
> ... representatives from governmental agencies,
> ... delegates of firefighter associations and organizations, and
> ... engineers from the fire insurance industry.

This unique feature provides a close relationship between the International Fire Service Training Association and other fire protection agencies, which helps to correlate the efforts of all concerned.

The publications of the International Fire Service Training Association are compatible with the National Fire Protection Association's Standard 1001, "Fire Fighter Professional Qualifications (1981)," and the International Association of Fire Fighters/International Association of Fire Chiefs "National Apprenticeship and Training Standards for the Fire Fighter." The standards are an effort to attain professional status through progressive training. The NFPA and IAFF/IAFC Standards were prepared in cooperation with the Joint Council of National Fire Service Organizations of which IFSTA is a member.

The International Fire Service Training Association meets each July at Oklahoma State University, Stillwater, Oklahoma. Fire Protection Publications at Oklahoma State University publishes all IFSTA training manuals and texts. This department is responsible to the executive board of the association. While most of the IFSTA training manuals can be used for self-instruction, they are best suited to group work under a qualified instructor.

Preface

The fire fighting profession has seen great changes over the past 20 years, not only in fire fighting equipment and methods, but also — and perhaps more importantly — in attitudes. For example, today's fireground strategies rely increasingly upon a more aggressive, interior fire attack and incorporate tactics that more effectively promote the firefighter's mission of preserving life and property. Modern forcible entry tactics naturally reflect this concern for life and property. While the indiscriminate destruction of building components was once the hallmark of forcible entry, today's tactics take into account the degree of damage to the structure involved.

This edition of **Forcible Entry** reflects the fire service's growing concern for the reduction of property damage, especially for the damage needlessly caused during fire fighting activities. Forcible entry does *not* have to result in heavy damage to building components. Forcible entry *can* be done quickly, and with minimal damage. The key to safe, efficient forcible entry lies in placing the proper tools into the hands of well-trained firefighters. This manual promotes that type of training.

As with any comprehensive training manual, a tremendous amount of work is required to bring up-to-date technical information and fireground-tested methods together into a format that is easy to read and adaptable to the greatest number of fire department operations. With grateful acknowledgement for the many hours of volunteer work that have been devoted to this manual, the following individuals are recognized:

Validating Committee

Chairman
Richard McIntyre, Retired
North Carolina State Training
Burlington, NC

Secretary
David Allen Cox
Asst. Training Specialist
Fire Protection Training Division
Texas A & M University
College Station, TX

Members

John M. Eversole
Lieutenant
Chicago Fire Department
Chicago, IL

Garland W. Fulbright
Executive Director
Commission on Fire Protection
    Personnel Standards and Education
Austin, TX

Bob Hasbrook
Captain
Neosho Fire Department
Neosho, MO

Norman J. Lewis
Training Officer
Daytona Beach Fire Department
Daytona Beach, FL

George Luther
Administrator
Connecticut State Fire Administration
Meriden, CT

John McNeece
Sergeant
Hayattstown Fire Department
Montgomery County, MD

David G. Moltrup
Lieutenant
Bethesda Fire Department
Montgomery County, MD

Steve Nuyen
Captain
Boulder City Fire Department
Boulder City, NV

Bob Ramirez
Assistant Chief
Los Angeles City Fire Department
Los Angeles, CA

Lloyd Scholer
Fire Instructors Association of Minnesota
Lake City, MN

Special thanks to the following departments and personnel for aid in staging and photographing pictures, as well as for technical assistance:

— Bob Rose of the Chico (CA) Fire Department, whose time and attention to photographic detail are much appreciated, as well as the "A" and "B" shifts at Station 2. Thanks, also, to Fire Chief Chuc Lowden and Battalion Chief Gerry Watson for their support.

— John McNeece, Dave Moltrup, and Joel Wood for supplying a major number of photos, as well as for providing technical assistance for through-the-lock entry techniques.

— Ralph Cooper, Stillwater, Oklahoma, for technical assistance on locks.

— Tom Brennan, Editor of *Fire Engineering* magazine, and Paul McFadden, Lt. John Cerato, Lt. John T. Vigiano of the New York City Fire Department for technical assistance and support. Thanks also to Capt. Gerry Grey, Lt. Chuck Krieger, Capt. Dan Kiely, and B/C Frank Blackburn for technical assistance.

— R. Thurston, El Paso, Texas, for writing review questions and answers.

And finally, our gratitude is extended to the following Fire Protection Publications staff members, whose time and talent made final publication of this manual possible:

David W. England, Senior Publications Editor
Lynne C. Murnane, Senior Publications Editor
Suzanne Goodwin, Technical Writer
Carol Smith, Publications Specialist
Don Davis, Coordinator, Publications Production
Ann Moffat, Graphic Designer
Mike McDonald, Graphic Artist
Desa Porter, Phototypesetter Operator II
Karen Murphy, Phototypesetter Operator II
Cindy Brakhage, Unit Assistant
Robert Fleischner, Photographer
Scott Stookey, Research Technician
Roger McKim, Research Technician
Kevin Roche, Research Technician
Robert Price, Research Technician
Michael Buchholz, Assistant Professor

Gene P. Carlson
Editor

Glossary

A

Adz — A tapered blade on a forcible entry tool, used for prying.

Astragal — A molding that covers the narrow opening between adjacent double doors in the closed position.

Auxiliary Deadbolt — A deadbolt bored lock; also called "tubular deadbolt."

Auxiliary Lock — A lock added to a door to increase security.

Awning Window — A type of swinging window that is hinged at the top and swings outward, often having two or more sections.

B

Backdraft — An explosion caused by an extremely rapid ignition and burning of heated gases within a confined area, usually after a sudden inrush of air, as when a door is opened; the degree of explosive force depends on such variables as the amount and speed of air that reaches the fire area, the amount of heat in the area, and the amount of gaseous fuel present.

Back Plate — The plate used with a rim lock to secure the lock cylinder to the door.

Balloon-Frame Construction — A type of wood-frame construction in which exterior studs are continuous from the foundation to the roof.

Battering — The act of creating an opening in a building component by striking and breaking it with a tool, such as a sledge or ram.

Bored Lock — A lock installed within right-angle holes bored in a door; also called "cylindrical lock."

Box Lock — A lock mortised into a door; also called "mortise lock."

Brace Lock — A rim lock equipped with a metal rod that serves as a brace against the door.

Breach — To break through, usually with the use of tools, a barrier such as a wall.

C

Cam — The part of a mortise lock cylinder that moves the bolt or latch as the key is turned.

Case (Lock) — The housing for any locking mechanism.

Casement Window — A type of swinging window that is hinged on the side and swings outward.

Curtain Door — A door used as a barrier to fire, consisting of interlocking steel plates or of a continuous formed spring steel "curtain"; curtain doors are often mounted in pairs, one door on the inside and the other on the outside of an opening.

Curtain Wall — An exterior non-load-bearing wall.

Cylinder (Lock) — The component of a locking mechanism that contains coded information for operating that lock, usually with a key.

Cylinder Guard — A metal plate that covers a lock cylinder to prevent forceful removal.

Cylinder Plug — The part of a lock cylinder that receives the key; also called the "key plug."

Cylinder Shell — The external case of a lock cylinder.

Cylindrical Lock — A lock having the lock cylinder contained in the knob; also called the "bored lock."

D

Deadbolt — The movable part of a deadbolt lock that extends from the lock mechanism into the door frame to secure the door in a locked position.

Dead Latch — A sliding pin or plunger that operates as part of a dead locking latch bolt; also called "anti-shim device."

Detention Window — A window designed to prevent exit by the occupants through a window opening.

Drop Bar — A metal or wooden bar that serves as a locking device when placed or "dropped" into brackets across an in-swinging door.

E

Exit Device — A locking assembly designed for panic exiting that unlocks from the inside when a release mechanism is pushed; also called "panic hardware."

F

Fire Door — A door, usually of metal, made to resist the passage of fire through an opening; operates by swinging or sliding from the side, or rolling down from overhead.

Fire Stop — Solid material, such as wood block, placed within a wall void to retard or prevent the spread of fire through the void.

Fire Wall — A wall designed to withstand severe fire exposure and to act as an absolute barrier against the spread of fire.

Fixed Window — A window that is set in a fixed or immovable position and cannot be opened for ventilation.

Flashover — The condition of fire in an enclosed space in which the entire atmosphere within the space becomes so hot that it suddenly becomes totally engulfed in flame.

Flush Bolt — A locking bolt that is installed flush within a door.

Folding Door — A door that opens and closes by folding; also called, if the door folds into several sections, an "accordion door."

Force — To break open, into, or through.

Forcible Entry — The techniques used to get into buildings or other areas of confinement when normal means of entry are locked or blocked.

Frame — The part of an opening that is constructed to support the component that closes and secures the opening, such as a door or window; also called "jamb."

Fulcrum — The support or point of support on which a lever turns in raising or moving something.

G

Glass Door — A door consisting primarily of glass, usually set in a metal frame.

Glazing — The part of a window that allows light to pass; the glass or thermoplastic panel in a window.

H

Handtool — A tool that is manipulated and powered by human force.

Hasp — A fastening device consisting of a loop eye or staple and a slotted hinge or bar; commonly used with a padlock.

Hinged Door — A swinging door.

Hopper Window — A type of swinging window that is hinged at the bottom and swings inward.

I

In-Swinging Door — A door that swings away from someone who stands on the outside of an opening.

J

Jalousie Window — A type of swinging window, usually with small louvered glass sections that open and close by turning a crank.

Jamb — The part of an opening that is constructed to support the component that closes and secures the opening, such as a door or window; also called "frame."

Jimmy — To pry apart, usually to separate the door from its frame to allow the latch or bolt to clear its strike.

Jimmy-Resistant Lock — An auxiliary lock having a bolt that interlocks with its strike and thus resists prying; also called "vertical deadbolt" or "interlocking deadbolt."

K

Kalamein Door — A metal-clad door.

Key — A device that, when inserted into a key plug, causes the internal pins or disks to align in a manner that allows the plug to turn within the cylinder; a device that allows the operator to lock and unlock a locking mechanism.

Key Box — A boxlike container that contains keys to the building, usually mounted on or in the front wall; requires a master key to open.

Key-in-Knob Lock — A lock in which the lock cylinder is within the knob.

Key Plug — The part of a lock cylinder that receives the key; also called the "cylinder plug."

Key Tool — A tool for manipulating an exposed lock mechanism so that the latch or deadbolt is retracted from its strike.

Keyway — The opening in a cylinder plug that receives the key.

L

Laminated Glass — Glass made of layers of glazing bonded to sheets of plastic; also called "safety glass."

Latch — The spring/loaded part of a locking mechanism that extends into a strike within the door frame.

Latch Bolt (Dead-Locking) — A latch with a shim or plunger that causes the latch to operate in a manner similar to a deadbolt; the latch plunger prevents "loiding" of the latch.

Ledge Door — A door constructed of individual boards joined within a frame; also called "batten door."

Lever — A device consisting of a bar turning about a fixed point (fulcrum), using power or force applied at a second point to lift or sustain an object at a third point.

Leverage — The action or mechanical power of a lever.

Load-Bearing Wall — A wall that is used for structural support.

Lock — A device for fastening, joining, or engaging two or more objects, such as a door and frame, together.

Lock Mechanism — The moving parts of a lock, which include the latch or bolt, lock cylinder, and articulating components.

Loiding — The method of slipping or shimming a spring latch from its strike with a piece of celluloid (credit card).

M

Mechanical Advantage — A gain in force when levering by moving the fulcrum closer to the object.

Metal-Clad Door — A door with a metal exterior; may be flush-type or panel-type; also called "Kalamein door."

Mortise — A notch, hole, or space cut into a door to receive a lock case, which contains the lock mechanism.

Mortise Cylinder — A lock cylinder for a mortise lock.

Mortise Lock — A lock mortised into a door; also called "box lock."

Mullion — A center post, sometimes removable, in a double door opening.

Multibolt Lock — A high-security lock that uses metal rods to secure the door on all sides.

N

Night Latch — A button on a rim lock that prevents retracting the latch from the outside.

O

Out-Swinging Door — A door that swings toward someone who stands on the outside of an opening.

Overhead Door — A door that opens and closes above a large opening, such as in a warehouse or garage; usually of the rolling, hinged panel, or slab type.

P

Padlock — A detachable, portable lock with a hinged or sliding shackle.

Panel Door — A door inset with panels, which are usually of wood, metal, glass, or plastic.

Panic Hardware — A locking assembly designed for panic exiting that unlocks from the inside when a release mechanism is pushed; also called "exit device."

Parapet — A low wall, usually at the edge of a roof, that extends above the roof line; often an extension of a fire wall, designed to prevent the spread of fire above the roof line.

Patio Door — A sliding glass door, commonly placed in an opening that accesses the patio or rear of a residence.

Pivoting Deadbolt — A lock having a deadbolt that pivots 90 degrees, designed to fit a narrow-stiled door.

Pivoting Window — A window that opens and closes either horizontally or vertically on pivoting hardware.

Plate Glass — Ground and polished, clear sheet glass; also called "float glass," although float glass is made by a different process.

Platform-Frame Construction — A type of wood-frame construction in which exterior studs run from the floor to the ceiling of each story.

Power Tool — A tool that acquires its power from a mechanical device, such as a motor or pump.

Preassembled Lock — A lock designed to be installed as a complete unit (requiring no assembly) within a door; also called "unit lock."

Prefire Plan — A report consisting of information about a building and the business(es) within the building; usually includes layout drawings of the site and building, as well as essential information about building features, hazards, and other facts that are of strategic importance in case of a fire or emergency incident.

Projected Window — A type of swinging window that is hinged at the top and swings either outward or inward.

Pry — To raise, move, or force with a prying tool.

R

Rail — A horizontal member of a window sash.

Revolving Door — A door made of three or four sections, or wings, arranged on a central pivot, that operates by rotating within a cylindrical housing.

Rim Cylinder — A lock cylinder for a rim lock.

Rim Lock — A type of auxiliary lock mounted on the surface of a door.

S

Safety Glass — Laminated glass.

Sash — The framework of a window, made of vertical stiles and horizontal rails, that holds the glazing.

Security Window — A window designed to prevent illegal entrance to a building through a window opening.

Shackle — The hinged part of a padlock.

Shear Line — The space between the shell and the plug of a lock cylinder obstructed by tumblers in the locked position.

Shove Knife — A tool for "loiding" a latch.

Skeleton Key — A key for a warded lock.

Stem — The part of a lock cylinder that activates the bolt or latch as the key is turned; also called "tailpiece."

Stile — A vertical member of a window sash.

Strike — The metal plate mounted in the door frame that receives the latch or deadbolt.

Slab Door — A door that has the appearance of being made of a single piece, or "slab" of wood; may be of two types — either hollow core or solid core.

Sliding Door — A door that opens and closes by sliding, usually on rollers, across its opening.

Surface Bolt — A sliding bolt installed on the surface of a door.

Swinging Door — A door that opens and closes by swinging from one side of its opening, usually on hinges; also called "hinged door."

T

Tailpiece — The part of a lock cylinder that activates the bolt or latch as the key is turned; also called "stem."

Tempered Glass — A glass made resistant to breaking by a special heating process known as "tempering."

Thermoplastic Glazing — A plastic glazing made of acrylic, butyrate, or polycarbonate plastic, and known for its resistance to breakage.

Thumbturn — A part of the lock, other than the key or knob, used to lock and unlock the door.

Tilt-Up Wall — A precast concrete wall that is raised or tipped up into position with a crane.

Tin-Clad Door — Similar to a metal-clad door, except covered with a lighter-gage metal.

Tubular Deadbolt — A deadbolting bored lock; also called "auxiliary deadbolt."

Tumbler — A pin in the tumbler-type of lock cylinder.

U

Unit Lock — A lock designed to be installed in a cut-out within the door without requiring disassembly and reassembly of the lock; also called "preassembled lock."

V

Veneer — A surface layer of attractive material laid over a base of common material, as with a veneered wall (faced with brick), or a veneered door (faced with a thin layer of hardwood).

W

Warded Lock — A simple type of mortise lock that requires a skeleton key to open.

Wired Glass — Sheet glass containing wire netting, which increases resistance to breakage and penetration.

Introduction

The threat of fire to life and property has been a perplexing problem since mankind first discovered fire. Protection from uncontrolled fire is an essential service provided by today's firefighters, whether they work as volunteers or on a full-time basis. In either case, the successful control and suppression of fire depends upon the firefighters' ability to quickly evaluate each fire situation and to react in an appropriate manner. This reaction to threatening situations involves the efficient, effective deployment of the fire fighting force to eliminate the threat as quickly as possible.

Because protection of life is the most important objective in any fire fighting operation, gaining entrance to a building is an essential tactical aspect of fire fighting strategy. Interior access is essential to searching the premises for occupants, as well as to providing the base for a strong interior attack of the fire.

Firefighters often find locked doors or other barriers blocking access into a building and to its occupants. For successful rescue and suppression to take place, such obstructions must be effectively removed, with speed and without unnecessary damage. For example, firefighters arriving at a building might see flames and smoke billowing from windows or the roof. Basic fire attack strategy calls for finding the seat of the fire rather than automatically directing hose streams through doors and windows and perhaps pushing the fire towards unburned areas. Gaining entry, then, is one of the operations vital to effective fire fighting. Ideally, entry should be as easy as turning a doorknob, but this is seldom the case. When entry is blocked, firefighters must use proper tools and imaginative *forcible entry* techniques.

Broadly defined, *forcible entry* means the techniques used to get into buildings or other areas of confinement when normal means of entry are locked or blocked. As building construction and security techniques improve, the methods required to force entry must also change. The trend today is toward stronger and more numerous locking mechanisms designed to keep out intruders. As lock design becomes more complex and door and window construction becomes more resistant to forcible entry, the firefighter's job also becomes more complex and demanding (Figure I.1).

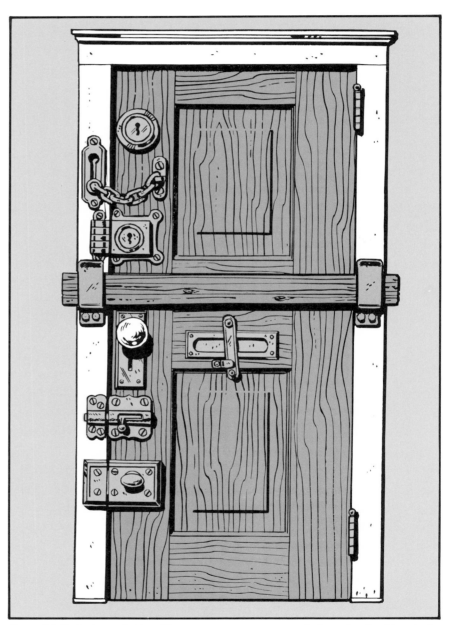

Figure I.1 Today's trend of installing stronger, more numerous locks makes forcible entry increasingly difficult.

Most modern architectural design incorporates building features that not only provide security from unwelcome intrusion, but also environmental protection for energy conservation. To this end, modern doors and windows are made of exceptionally strong materials, placed within frames of similar strength, and secured with locks of sophisticated design. As technology changes, so forcible entry training must change to reflect the latest in building construction and design. The fire service must stay abreast of these changes in order to develop forcible entry techniques that facilitate entrance into even the most well-built structures.

The efficiency of firefighters in forcing entrance to a building is dependent on two major factors: (1) choosing the appropriate tool(s) and (2) applying the proper technique to the structural component. Forcing entry when standard openings are locked means using special skills to open doors and windows with various types of tools. Equipment may vary from simple hand tools, such as axes and pry bars, to more specialized tools, such as hydraulic spreading equipment or air bags. Skill with forcible entry tools comes primarily through training, preferably hands-on training on buildings that present a variety of forcible entry problems. By confronting various situations during training exercises, firefighters can develop a number of strategies for forcing entrance. Training exercises also present opportunities to encourage innovation in approaches to each problem. Forcible entry requires that individuals not only be able to apply basic principles, but also that they be ready to adapt established methods to accommodate the problem at hand.

The efficiency of forcible entry operations involves more than the successful penetration of a building exterior. Entry must also be accomplished with a minimum of damage. Breaking a lock, prying apart a jamb, or forcing a window may cause some damage, but the damage is generally far less than if firefighters resort to the indiscriminate destruction of doors and frames or large sections of window glass (Figure I.2). The ability to gain entrance with a minimum of damage is a forcible entry objective consistent with the overall goal of preserving property. An efficient forcing operation also, by definition, should be accomplished within a relatively short time. Fire fighting and rescue success depends upon gaining entrance with as little delay as possible. Fire moves with remarkable speed when conditions of fuel, heat, and air are optimum, and victims within a fire-involved building can be overcome quickly by toxic products of combustion (Figure I.3).

The method used to force a door depends upon its construction, mounting, and locking mechanism. Likewise, forcing entrance through a window involves an assessment of its construction and operational design, as well as identification of its locking mechanism. A knowledge of major design features common to most doors and windows is essential to determining the best means to gain entrance quickly with the least amount of damage.

Forcing entry when normal openings are blocked may also involve removing external obstructions, such as steel bars (Figure I.4), heavy wire-mesh guards, or metal shutters, as well as internal obstructions such as piles of stock or debris.

Many of today's buildings either have a minimal number of openings or have doors and windows that are exceptionally resistant to forcing. Firefighters, therefore, should also be trained in basic building construction principles. This training will help them to recognize parts of structures that are most vulnerable to forcible entry.

Figure I.2 Indiscriminate damage during forcible entry violates one of our primary missions: PROTECTION OF PROPERTY.

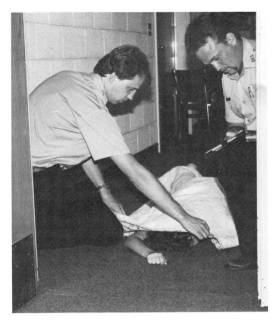

Figure I.3 Many fire victims have died just seconds before being found. Could they have been saved if forcible entry had been done more efficiently?

Figure I.4 Steel bars are not only obstacles to forcible entry crews; they are also deadly barriers to victims within. *Courtesy of Edward Prendergast.*

PURPOSE AND SCOPE

Skillful forcible entry saves lives, lessens property damage, and improves public relations. This manual is designed to help firefighters develop the knowledge and skills to perform forcible entry efficiently. It can be used by instructors to teach firefighters how to solve forcible entry problems on the drill ground and, ultimately, on the scene of an emergency. The text addresses the standards for Firefighter I, II, and III, as set forth in the NFPA Standard 1001, *Fire Fighter Professional Qualifications*.

To prepare the firefighter to deal with the problems typically encountered in forcible entry situations, this manual covers both the construction features and forcible entry techniques for basic door and window assemblies, lock assemblies, and walls. The manual also describes the tools used for performing forcible entry, as well as the way they work. In addition, it describes specific forcible entry problems, and offers suggestions for dealing with these unusual situations.

Principles of
Forcible
Entry

1

NFPA STANDARD 1001
FORCIBLE ENTRY
Fire Fighter I

3-2 Forcible Entry

3-2.1 The fire fighter shall identify and demonstrate the use of each type of manual forcible entry tool.

3-2.2 The fire fighter shall identify the method and procedure of properly cleaning, maintaining, and inspecting each type of forcible entry tool and equipment.

Fire Fighter II

4-2 Forcible Entry

4-2.1 The fire fighter shall identify materials and construction features of doors, windows, roofs, floors, and vertical barriers and shall define the dangers associated with each in an emergency situation.

4-2.2 The fire fighter shall identify the method and technique of forcible entry through any door, window, ceiling, roof, floor, or vertical barrier.

Chapter 1
Principles of
Forcible Entry

When firefighters approach a potentially life-threatening or property-threatening situation, they must decide whether a defensive or offensive strategy is appropriate. On those rare occasions in which a fire has hopelessly engulfed a major portion of the structure, a defensive position is often taken. In this case, the incident commander has chosen to attack the fire from the exterior, which means protecting exposures and containing the fire as much as possible from a position outside the structure. Because the fire attack is made from outside the structure, forcible entry is usually not required(Figure 1.1). A defensive strategy may be the

Figure 1.1 An exterior attack on a fire usually does not require forcible entry. *Courtesy of Rick Bellatti, Stillwater (OK) News Press.*

most realistic approach to dealing with a fire that has exceeded the capabilities and resources of the fire department. It may also be the most practical approach to preventing unnecessary risks to fire fighting personnel. The fire department is responsible for protecting life and property, and this responsibility must extend to and include the fire fighting force itself.

An aggressive fire fighting force chooses the offensive mode of operation whenever possible. Gaining interior access to the structure is an intermediate objective when attacking a fire offensively. Concentrated, effective action to control the situation can be taken only from an interior position (Figure 1.2). More importantly, the rescue of victims can be satisfactorily accomplished *only* from the interior. For these reasons, forcible entry is of major importance in achieving overall offensive objectives.

Figure 1.2 Attacking a fire from the interior requires that forcible entry be made quickly and efficiently. *Courtesy of Mike Wieder.*

An efficient forcible entry operation must not only provide access to the building interior, but it must also be performed with speed and with a minimum of damage. At times it may be necessary to sacrifice the objective of minimizing damage to enter the building with the greatest speed. An expensive door, for example, may have to be battered open to provide the shortest, fastest route for access and removal of a victim.

It should be remembered, however, that the majority of incidents do not justify indiscriminate damage in order to deal with the problem. A majority of a fire department's activity may be "nuisance" calls, such as malfunctioning fire alarm devices and false alarms. In such instances, forcible entry may still be warranted in order to enter and investigate a locked building to assess whether an emergency condition exists (Figure 1.3). In such

Figure 1.3 Forcible entry is often required when investigating "nuisance" calls; in these situations, damage must be minimized. *Courtesy of Loren Dunlap.*

situations, it is necessary to force entry with an absolute minimum of damage.

Public opinion about the fire department's action at nuisance calls is part of the reason that close attention must be paid to methods of forcing entrance, especially when no visible signs of an emergency exist. The old saying, "A ten-cent fire and the fire department did a thousand dollars worth of damage," may have come about, at least in part, because of indiscriminate forcible entry practices. Too often, overzealous firefighters make entrance with only one criteria in mind: speed. When this happens, unnecessary damage usually occurs.

It was once thought that destruction was a necessary cost for forcing entrance to a locked building in which a fire was thought to exist. It was accepted that for forcible entry to be made effectively, more than minor damage was inevitable. We now realize that major damage is not a necessary part of forcible entry. Damage should decrease as the knowledge and skill of entry crews increase. It follows, therefore, that training is an integral part of an effective forcible entry operation.

TRAINING

Training in solid forcible entry techniques should include a number of elements:

- Tool design and use

- Door, window, and lock construction
- Basic building construction
- Forcible entry methods and techniques
- Safety during forcible entry
- Pre-incident planning

Tool Design and Use

Standard forcible entry equipment on apparatus once consisted of no more than a few axes and crow bars. As attitudes about forcible entry changed, that is to say, as speed *and* minimum damage came to be considered as necessary functions of effective forcible entry, more specialized tools were developed (Figure 1.4). Training in proper tool usage is an integral part of learning forcible entry techniques. To learn how a tool works, an understanding of how force is transmitted through the tool and multiplied through leverage is necessary. This knowledge gives the user a better idea about the capabilities and limitations of each tool and facilitates the selection of the best tool for each situation.

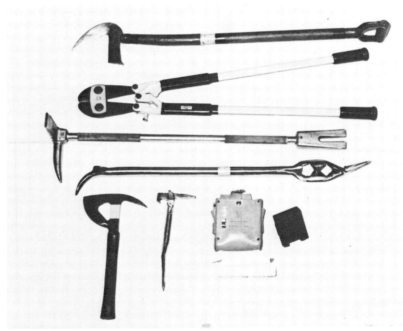

Figure 1.4 Specialized tools make it possible to carry out forcible entry quickly, with a minimum of damage.

Door, Window, and Lock Construction

An understanding of door and window construction is a prerequisite for determining the best means of entry through building openings. From the standpoint of damage control, a knowledge of how basic locking devices operate is also essential because locks often provide the best means of entrance with the least amount of damage (Figure 1.5). Familiarity with the details of each basic type of door, window, and lock assembly will allow you to identify those components that most readily yield to forcing actions.

Basic Building Construction

Buildings designed with few openings present a significant forcible entry problem. For this reason, it is essential that firefighters possess a basic knowledge of building construction as an aid to determining the most accessible points through which a building can be entered when openings cannot be utilized. Without an understanding, for example, of how breaching a wall will affect a building's structural integrity, breaching a wall is not a feasible tactic. Forcible entry through walls can be a time-consuming, inefficient method of forcible entry if firefighters are not clear about which building components yield most readily to breaching operations.

Forcible Entry Methods and Techniques

There are a number of tested and reliable methods for forcing entry through doors and windows, including those that permit entrance through locking mechanisms. These methods are based upon basic types of door, window, and lock designs. When new or unique designs are encountered, standard forcible entry methods must be adapted to accomplish entry. In these instances, the imagination and innovation of the firefighter are essential to forcible entry. Common sense is by far the best approach to solving such problems. While "standard operating procedures" may provide guidelines for most situations, the individual firefighter still must make decisions about the best point of entry, tool(s), and method to accomplish entry.

Safety During Forcible Entry Operations

Job safety should be included in every segment of forcible entry training. Safety is one of the underlying reasons for learning the basics of building construction: by understanding how a building is assembled, firefighters should become aware of potential areas in which careless breaching could cause structural failure. By understanding how basic types of doors, windows, and locks are made, the firefighter can choose the safest method for forcing these devices. It is also essential that personnel understand the proper use of tools so that safety rules are not violated. Finally, the use of protective equipment, such as gloves and eye protection, should be enforced during all training exercises, as well as during actual incidents.

Pre-Incident Planning

One of the best ways to solve forcible entry problems is to identify them *before* an incident. The only realistic way to accom-

Figure 1.5 Through-the-lock entry often provides the least-destructive means of forcing entrance. This door was forced open without damage by pulling the lock cylinder.

Figure 1.6 Doors, windows, and locks should be identified and noted during pre-incident surveys.

plish this is through a systematic "mapping" of buildings in each response area. While prefire planning is a routine exercise in many fire departments, potential forcible entry problems are often overlooked. Building openings are usually located and identified on prefire survey schematics, but little is done to pinpoint specific potential entry problems.

Special attention should be given to door and window designs, as well as to the locking mechanisms within each building (Figure 1.6). Once this is accomplished, optimum points of entry can be noted on the prefire plan. More than one point of entry should be identified in case of unforeseen problems (Figure 1.7). Identifying the best entry points should be based upon the following considerations:

- Feasibility of rescue and fire suppression tactics
- Vulnerability of assemblies to forcible entry methods
- Time required to make entry
- Potential for damage
- Accessibility

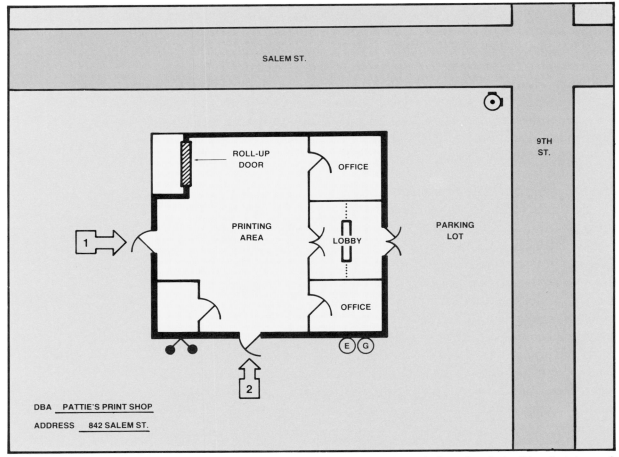

Figure 1.7 Prefire survey drawings should indicate the best place for forcible entry in terms of accessibility, ease of entry, and potential for damage. At least one alternate entry point should also be indicated.

Another benefit of pre-incident planning is that crews can discuss potential problems with building owners and managers. Unique problems like guard dogs and special property barriers can be identified and noted on survey forms (Chapter 10 discusses methods to solve these special problems).

It may be possible to make arrangements for nondestructive means of entering the building. A primary method of nondestructive entry to a locked building is, of course, with a key. Because it is not practical for the fire department to carry keys to all buildings in its jurisdiction, key boxes can be provided on the outside of buildings (Figure 1.8). A prefire inspection offers an excellent opportunity for the fire department to educate owners and managers about the availability of such devices as a means of preventing forcible entry damage in case of fire.

Figure 1.8 Exterior key boxes provide a means of nondestructive entry. *Courtesy of The Knox Company, Newport Beach, CA.*

There is no substitute for training as a means to ensure that forcible entry is made in the most efficient manner. Efficient forcible entry is speedy, minimizes damage, and is accomplished without injury to personnel. It is a critical phase of an offensive fire attack and must be carried out with little delay so that an aggressive interior fire attack and rescue operation can take place. Training provides not only the opportunity to sharpen skills, but also enables firefighters to accept forcible entry assignments with confidence and reasonable assurance that the job will be carried out successfully.

Answers on page 261

Complete the following statement with words or phrases that make the statement correct:

1. For a forcible entry operation to be deemed "efficient" it must not only provide access to the building interior but must be accomplished _____ and _____.

Answer each of the following questions in a few words or short phrases:

2. Points of entry to a building should be identified during prefire planning inspections. What four criteria should be used to identify the best entry points?

   A. _____
   _____

   B. _____
   _____

   C. _____
   _____

   D. _____
   _____

3. Why is speed so important to forcible entry operations on the fireground?
   _____
   _____

Determine whether the following statements are true or false. If false, state why:

4. Forcible entry means the techniques used to get into buildings or other areas of confinement when normal means of entry are locked or blocked.
   ☐ T   ☐ F   _____
   _____

5. Skill in utilizing forcible entry tools comes primarily through classroom training on tool design and use.
   ☐ T   ☐ F   _____
   _____

6. For forcible entry to be made effectively, more than minor damage is inevitable.
   ☐ T   ☐ F   _____
   _____

7. When new and unique designs in door, window, and lock construction are encountered during prefire inspections, new forcible entry techniques must often be developed and practiced because standard methods will not work.

☐ T ☐ F _____

_____

8. Prefire plans for a building should identify more than one point of entry to the building.

☐ T ☐ F _____

_____

Select the choice that best completes the sentence or answers the question:

9. Forcible entry training should include
A. tool design and use
B. basic building construction
C. lock construction
D. all of the above

10. What should be emphasized over all other aspects during forcible entry training, in both classroom and training exercises?
A. Speed
B. Safety
C. Common sense
D. Innovation

Forcible Entry Tools

2

**NFPA STANDARD 1001
FORCIBLE ENTRY
Fire Fighter I**

3-2 Forcible Entry

3-2.1 The fire fighter shall identify and demonstrate the use of each type of manual forcible entry tool.

3-2.2 The fire fighter shall identify the method and procedure of properly cleaning, maintaining, and inspecting each type of forcible entry tool and equipment.

Chapter 2
Forcible Entry Tools

A number of tools designed specifically for forcible entry are available today. Many tools are appropriate for a variety of forcible entry situations and are therefore more widely known and distributed. Other tools are very specialized and suit only specific applications in a few localities. Such tools have often been developed by firefighters who, through experience and experimentation, have devised tools for the specific forcible entry problems in their respective jurisdictions. These tools may be less well-known and difficult to find unless they are well-advertised and marketed.

Manufacturers are developing new tools each year to help prepare fire departments for an increasing number of unique forcible entry problems. New construction in every community inevitably produces an array of new forcible entry challenges that must be met by firefighters. A prefire planning system that identifies the potential forcible entry problems in both old and new construction is essential to prepare personnel to meet these challenges. If methods cannot be devised to deal with these problems using the tools at hand, it may be necessary to purchase new tools and to adopt new methods.

This manual discusses many well-known tools primarily because most departments are currently using these tools and will continue to do so. The most common tools are sufficient to meet a majority of forcible entry situations. This chapter also contains a discussion of different versions of some tools, as well as of some specialized tools.

Forcible entry tools can be divided into four groups, based on the manner in which they are used to force entry:

- Prying and spreading tools
- Cutting and boring tools

- Striking and battering tools
- Lock-entry tools

Tools from each division may also function in other categories. An axe, for example, can be used for both cutting and prying. Some tools are used in conjunction with others and thus can also be subcategorized. The following tools are therefore listed according to their *primary* or most common use.

Hand tools can be generally defined as tools that rely on *human force* to transmit power directly to the working end of the tool. It should be noted that many forcible entry hand tools can be used not only for prying and spreading, but also for cutting, striking, or even lock-entry work. These are often referred to as "multipurpose" or "utility" tools.

The use of power tools in forcible entry makes the job much easier than working with hand tools. A primary advantage of power tools is that they can generate over 20,000 psi (137,900 kPa) of force. There are four major divisions, classified by the medium used to generate force:

- Electric
- Hydraulic
- Pneumatic
- Pneumo-hydraulic

Electrically powered tools convert electrical energy through an electrical motor into mechanical energy. Hydraulic tools have compressor pumps that transmit force through a liquid (hydraulic fluid) to the working end of the tool. Hydraulic pumps are either hand operated or are powered by a gas or electric motor. Pneumatic tools use air pressure to transmit force to the working end of the tool. Pneumo-hydraulic devices combine both air and liquid force in an air-driven hydraulic pump that generates power to the forcing tool.

PRYING AND SPREADING TOOLS

Prying and spreading tools operate on the principle that if structural components are pushed or pulled out of alignment, they either break or lose integrity. Breaking a structural component usually causes the component to become functionally useless. Moving a component to the point that it loses integrity (for example, pushing a door frame out of plumb) can disable a locking device that depends upon the structural component as an integral part of the securing mechanism.

Prying hand tools use leverage to provide a mechanical advantage. This means that a person using a tool can generate more force to an object than if no tool were used. Many tools are designed to break locks, pry doors, and force windows. A tool that has a blade or an adz to penetrate a narrow opening, such as the space between a door and a frame, should be designed with a long, narrow taper. When the taper is six-to-one, that is, when the length of the head is six times its maximum thickness, the operator has the optimum mechanical advantage (Figure 2.1).

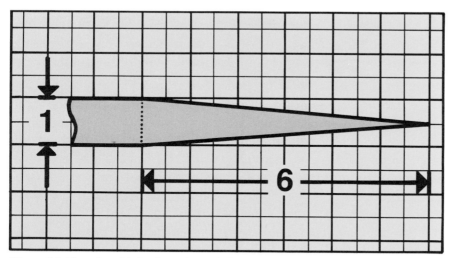

Figure 2.1 The adz or blade of a prying tool should be 6:1 for maximum mechanical advantage.

Another important principle in prying operations involves a basic principle of physics. When force is transmitted through a prying tool, the tool must be levered against a stationary object. When this object, the fulcrum, is located at the midpoint of the tool, the force transmitted to the objective is equal to the force applied at the opposite end (Figure 2.2). The closer the fulcrum is to the objective, the greater the mechanical advantage is to the operator. As the fulcrum is moved closer to the objective, mechanical advantage is increased (that is, force at the objective end is multiplied although force at the opposite end remains constant).

Mechanical advantage can be quickly calculated by comparing the tool length on each side of the fulcrum. For example, if the fulcrum is placed two-thirds of the length of the tool from the end at which force is applied (Figure 2.3), the mechanical advantage is two-to-one (2:1). This is because the length of tool between the

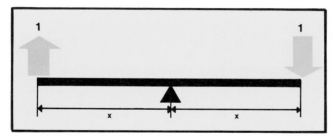

Figure 2.2 When the fulcrum of a prying tool is placed equidistant between the object pried and the applied force, the force transmitted to the object is exactly the same as the force applied.

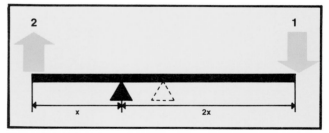

Figure 2.3 If the fulcrum is moved to a point one-third of the tool length from the object being pried, the mechanical advantage is doubled.

fulcrum and the force is twice that of the length between the fulcrum and the objective. Figure 2.4 illustrates how fulcrum placement is important in obtaining full advantage of the prying tool. A force of 100 pounds (45 kg) applied at the end of a pry bar can generate more than 1½ tons (1 361 kg) of force at the opposite end if the fulcrum is placed 1/32 of the length of the bar from the objective. In prying operations, place the fulcrum as close to the objective as possible to provide maximum mechanical advantage. Naturally, force should be applied as far away from the fulcrum as possible. This is why long-handled tools offer a significant mechanical advantage over shorter tools of identical design.

Varying the location of the fulcrum solves any number of forcible entry problems by maximizing the mechanical advantage of the prying tool. The same principle of mechanical advantage applies to a prying situation in which the objective is located at some point along the surface of the prying tool, as when a bar is used to break a padlock (Figure 2.5). In this case, the end of the tool acts as the fulcrum. The position of the fulcrum in relation to the padlock is critical to maximizing mechanical advantage. The closer the fulcrum is to the lock, the greater the advantage (Figure 2.6).

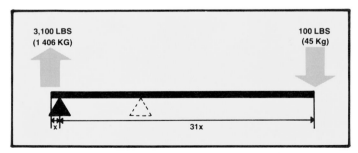

Figure 2.4 If the fulcrum is placed 1/32 of the tool length from the object, a 100-pound (45 kg) force is increased to 3,100 pounds (1 406 kg) of force at the object end.

Figure 2.5 No mechanical advantage is gained by positioning a pry bar so that the padlock is an equal distance from the fulcrum and the applied force.

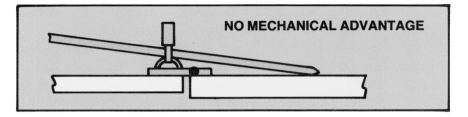

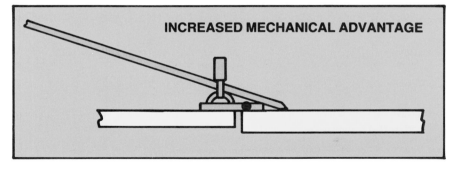

Figure 2.6 Positioning the padlock closer to the fulcrum end of the pry bar significantly increases mechanical advantage.

Air Bag

The air bag is a versatile device that is easily adaptable for forcible entry work. While it was originally designed for such uses as uprighting overturned tractor-trailer rigs or for rescue and extrication work, the air bag is also useful in forcible entry situations. It simply transmits the force of compressed air, usually supplied by compressed air bottles, throughout the surface of the bag. Although its pressures do not exceed 2,000 psi (13 790 kPa), this pressure is multiplied over every square inch of the bag. Clearly, a tremendous amount of power can be generated by the entire bag to a single object such as a door or barred window. An air bag inserted under a roll-up door, for instance, can force the door up enough to actually break locking bars (Figure 2.7). Skilled operators are limited only by their ingenuity in finding uses for this unique tool in situations that require creative application of the tools at hand.

Figure 2.7 An air bag can generate enough force to actually break the locking bars on a roll-up door. *Courtesy of Paratech Inc.*

Air bag systems have several distinct advantages over other types of lifting or spreading equipment. They have

- High Capacity — up to 146,800 pounds (66 588 kg) of lifting or spreading capacity.

- Wide Spreading Capability — up to 20 inches (508 mm). Two bags stacked together double this distance (Figure 2.8 on next page).

Figure 2.8 Two air bags in a stacked configuration can be used to spread up to 40 inches (1 016 mm). *Courtesy of Paratech Inc.*

● Narrow Space Requirement — A 1-inch (25 mm) space is all that is required to put the bag in place (Figure 2.9).

Figure 2.9 An air bag can be manipulated into a very narrow space. *Courtesy of Paratech Inc.*

● Quiet Operation — Air bags are powered by compressed air supplied from a standard self-contained breathing apparatus air cylinder (Figure 2.10) or other acceptable stored pressure vessel. It does not require any type of motorized power, thus it eliminates the noise, unreliability, and hazards usually associated with gasoline-powered rescue equipment.

Figure 2.10 Because they are powered by compressed air, one advantage of airbags is their quiet operation. *Courtesy of Paratech Inc.*

AIR BAG SAFETY RULES

Operators should follow these safety rules when using air bags (some rules may not necessarily apply to forcible entry operations):

● Plan the operation before starting the work.

● Be thoroughly familiar with the equipment: its operating principles, methods, and limitations.

● Keep all components in good operating condition and all safety seals in place.

● Have an adequate air supply and sufficient cribbing available before beginning operations.

● Position the bags on or against a solid surface.

● Never inflate the bags against sharp objects.

- Inflate the bags slowly and monitor them continually for any shifting.
- Never work under a load supported only by bags.
- Shore up the load with enough cribbing blocks to more than adequately support the load in case of bag failure (Figure 2.11).

Figure 2.11 Loads should be shored up after they are lifted by air bags. *Courtesy of Paratech Inc.*

- Stop the procedure frequently to increase shoring or cribbing.
- Be sure that the top layer is solid when using the box cribbing method: leaving a hole in the center may cause shifting and collapse (Figure 2.12).
- Avoid exposing bags to hot materials of more than 220°F (104°C): insulate the bags with a nonflammable material.
- Never stack more than two bags — center the bags with the smaller bag on top and inflate the bottom bag first (Figure 2.13).
- Consult individual operator's manuals and follow the recommendations for the specific system used.

Axe

The axe has been a basic fire service tool since fire brigades were first organized. The fire fighting axe is a tool that has many uses, but it is designed primarily for *cutting*. Two of the most popular designs are the flat head (Figure 2.14) and the pick head

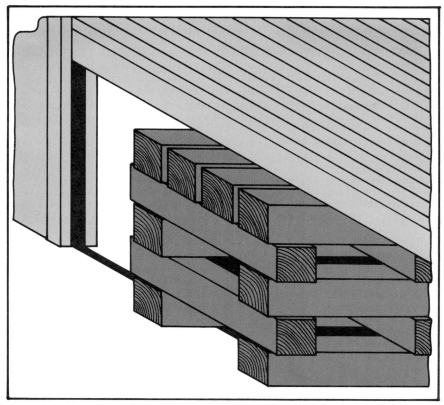

Figure 2.12 Make the top layer of box cribbing solid to prevent shifting and collapse.

Figure 2.13 When stacking bags of unequal size, place the smaller bag on top of the larger; inflate the bottom bag first. *Courtesy of Paratech Inc.*

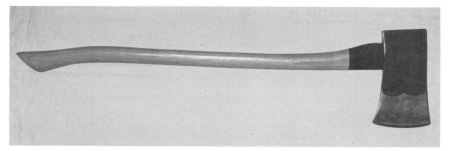

Figure 2.14. Although designed primarily for cutting, the flat-head axe is often used as a striking tool.

(Figure 2.15). The flat-head axe is often used as a striking tool, especially if used in conjunction with another tool during forcible entry. The pick-head axe can be used as a puncturing or picking device. Both types of axes have tapered blades that make them adaptable for prying, although handles tend to break when the head is twisted during a prying operation. This problem can be overcome by welding a steel handle to the head to create an unbreakable one-piece tool. The steel-handled axe should then be used primarily as a prying device, rather than as a cutting tool, because the axe is no longer properly balanced for swinging.

There is one type of axe designed for both cutting and prying. The pry-axe (Figure 2.16) is better for prying than the standard pick-head or flat-head axe. This tool is equipped with an extra handle that can be inserted into the head so that it can be turned at a 90 degree angle when prying. This is a better method of prying than using the twisting method done with a standard axe. The tool also incorporates other features such as a pick and a claw, which make it adaptable for other forcible entry methods.

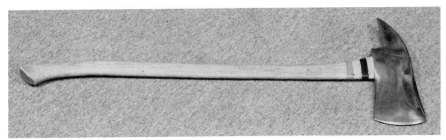

Figure 2.15 The pick-head axe can be used for prying, but there are other tools better suited for this purpose.

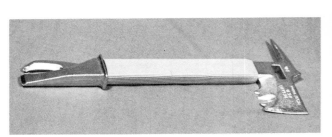
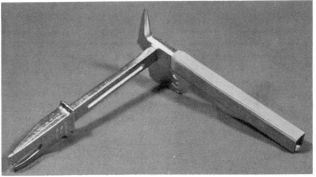

Figure 2.16 The pry-axe is designed for both cutting and prying. An extra handle (detachable) provides leverage for prying operations. *Courtesy of Paratech Inc.*

Claw Tool, Crowbar, Jimmy, Pry Bar, Wrecking Bar

The names of these tools and their features vary, depending upon the manufacturer. Constructed of iron or steel, each tool is available in lengths of up to 5 feet (1.5 m). The claw tool, designed for removing nails and small screws from wood, has a curved "claw" at one end and a split adz at the opposite end. Wrecking bars (also called crowbars) are designed to pull nails from wood and to dismantle building components. They are also well-suited for prying hasps, locks, and latches. Pry bars, jimmies, and crowbars are straight bars that have at least one beveled end. One pry bar model has an adjustable fulcrum built into the tool (Figure 2.17). As shown in Figure 2.18, these bars are also available in a variety of designs and lengths. A tool may be marketed under different names, depending upon the manufacturer, and as a result, may be identified by more than one name within a fire department.

Figure 2.17 This pry tool is especially designed for forcible entry. It has an adjustable fulcrum. *Courtesy of Gordon Harris, Elkhart Brass Mfg. Co., Inc.*

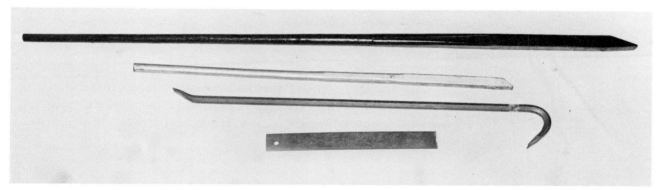

Figure 2.18 Pry bars, jimmies, crowbars, and wrecking bars are available in many varieties and sizes. *Courtesy of Bob Rose, Chico (CA) Fire Dept.*

Kelly Tool, Halligan Tool, Hooligan Tool, Hux Bar

These tools have features that make them more specialized for forcible entry work than simple pry bars or wrecking bars. The Kelly Tool (Figure 2.19) has two adz surfaces, one aligned with the handle (180 degrees) and the other at a right angle (90 degrees) to the handle. The Halligan tool, also known as the Hooligan (Figure 2.20) tool, resembles the Kelly tool but has a pointed pike for such uses as breaking padlocks and hasps. The pike can

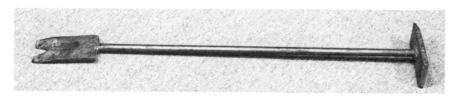

Figure 2.19 The Kelly tool has two adz surfaces, each at a different angle to the handle.

Figure 2.20 The Halligan tool (also called a Hooligan tool) can be used for many types of forcible entry operations.

either be levered or twisted to break the objective. The Hux Bar (Figure 2.21) is a variation of the Halligan tool and features pentagonal and square openings in the handle for operating hydrants.

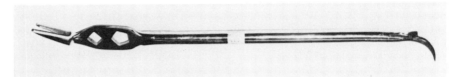

Figure 2.21 The Hux bar can be used for opening hydrants, as well as for a number of forcible entry tasks.

Lock Breaker

This tool is used to force open case-hardened locks that bolt cutters cannot cut. The lock breaker (Figure 2.22) has a triangular wedge (duckbill) that is used to spread a padlock apart. Inserting the wedge into the shackle of the lock and driving it in with a heavy hammer or sledge transmits spreading force sufficient to break the shackle.

Door Opener

The door opener is a compound lever that is used to force inward-swinging doors. This device, depending on manufacturing design, incorporates either a ratcheting mechanism or an articulating lever (Figure 2.23) to transmit force to the objective.

Figure 2.22 Few padlock shackles can withstand the spreading force of a duck-billed lock breaker.

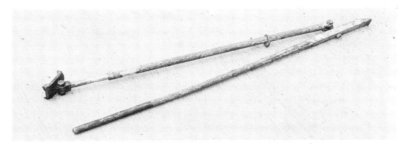

Figure 2.23 The Detroit door opener is one of the oldest forcible entry tools.

Spreader Bar

The spreader bar uses a screw yolk to transmit force equally to both ends of a bar. Force is applied by turning a handle located at a midpoint on the tool. This tool is useful for spreading door and window casings.

Miscellaneous Hand Tools

There are many other hand tools that can be used for prying and spreading. The simple firefighter's spanner (Figure 2.24) is an effective tool for forcing simple doors and windows. Because of its compactness, it can be carried in the turnout coat pocket or in a pouch or holder sewn to the turnout coat or pants.

Smaller tools like screwdrivers, claw hammers, center punches, and chisels should be included in every tool kit (Figure 2.25). These tools work well in some forcible entry situations. As firefighters become familiar with the features of doors, windows, and locks, they will realize the need for tools designed especially for specific applications. In some cases these tools can be designed and made in the fire station shop by fire department personnel.

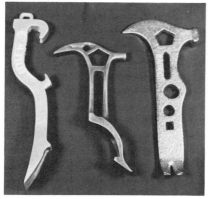

Figure 2.24 The firefighter personal spanner can be used for simple forcible entry operations.

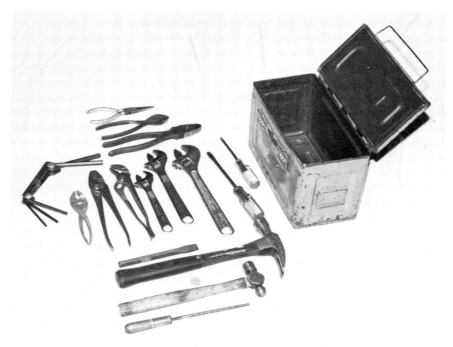

Figure 2.25 Every tool kit contains a variety of tools that can be adapted to forcible entry use.

Hydraulic Multipurpose Tool

The hydraulic multipurpose tool transmits pressure from a hand-pumped compressor through a hydraulic hose to a tool assembly (Figure 2.26 on next page). The device is labeled "multipurpose" because it can be applied to many types of problems. The tool was actually developed for rescue and extrication work, but it has excellent applications for forcible entry work as well. It comes

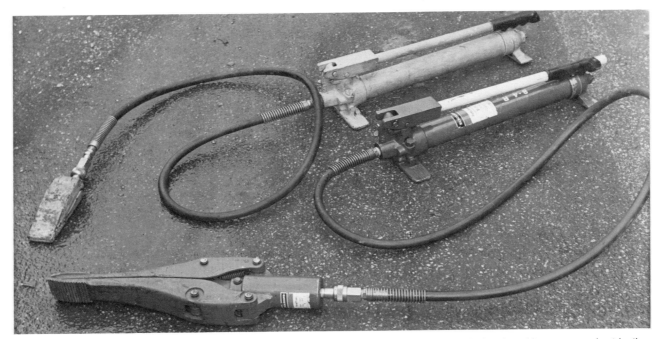

Figure 2.26 The hydraulic multipurpose tool, originally developed for rescue and extrication, is an excellent forcible entry tool.

with a number of accessories that, depending upon how they are combined, transmit force in different directions. For example, the tool can be used to *pull* an out-swinging door or to *spread* a sliding door. An advantage of this tool over the simple jack is that the spreading force can be initiated within a much smaller space than that required by a jack. The multipurpose tool is adaptable to many situations, although assembling complex combinations of accessories can be time consuming.

Another tool specifically designed for forcing doors and windows is the "rabbit" tool (Figure 2.27). The primary advantage of this type of device is that its working end, a pair of intermeshed spreading tips, can be inserted into a very narrow opening such as that between a door and its casing. A few strokes to the hand-

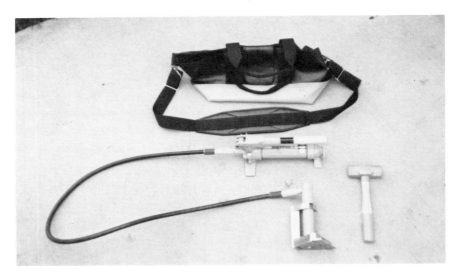

Figure 2.27 The rabbit tool is designed to open doors that swing inward. *Courtesy of Clemens Industries.*

pump compressor are usually sufficient to force open doors that swing inward.

Powered Hydraulic Multipurpose Tool

Powered hydraulic tools have pumps that are driven by either gas or electric motors (Figure 2.28), or by compressed air (Figure 2.29). The primary advantage of powered tools is speed. Like hand-pumped hydraulic tools, these devices are designed primarily for extrication work requiring speed and exceptional power to remove victims from vehicle wreckage or from machinery entanglement. Powered hydraulic tools are naturally suitable for many forcible entry situations. Doors that were once considered inpenetrable barriers to firefighters yield easily to the force of powered hydraulic spreaders. Metal barred windows that usually take several minutes to force with conventional hand tools can be opened in seconds with a skillfully used powered tool. Although it generally takes a few moments longer to set up the motorized equipment that drives these tools, the time is quickly regained once the forcing operation begins.

Figure 2.28 Powered hydraulic tools, with pumps driven by either gas or electric motors, are available with many accessories. *Courtesy of JAWS OF LIFE Rescue Systems, Hale Fire Pump Company.*

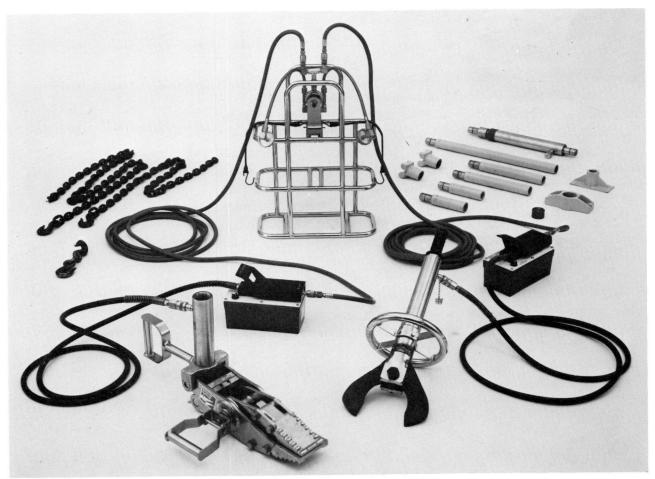

Figure 2.29 One type of hydraulic power tool uses an air-driven hydraulic pump. *Courtesy of Lincoln Safety Products.*

Power spreaders have other advantages over similar hand-pumped equipment. Forces of up to 20,000 psi (137 900 kPa) can be generated at the tips, which can spread up to 30 inches (762 mm). This is significant when selecting the proper tool for the situation is critical to quick entry. Some departments automatically take powered tools into any commercial or industrial incident with the attitude that a forcible entry situation usually warrants the use of such tools.

CUTTING AND BORING TOOLS

Another way to force entry is to cut structural components. Cutting tools, like prying and spreading tools, can be used to alter a structural member and cause it to lose its strength and integrity. When this alteration occurs, of course, entry can be made.

Cutting tools are designed to cut specific materials and should be used only for the purpose intended by the manufacturer. Failure of tools in entry operations is most often due to mismatching the tools with the material to be cut.

Cutting tools work in a variety of ways. One of the simplest tools, the axe, depends upon the force of velocity as well as upon the sharpness of its blade to penetrate material. Chisels work in a similar manner but rely upon a striking tool or upon pneumatic force to accomplish the same purpose. Bolt cutters use force levered through articulating handles to cut material by cleaving. Bolt cutter jaws are composed of extremely hard metal, which is necessary for penetrating metals such as those found in padlock hasps and window guard bars. Shears, both manual and hydraulic, rely upon jaws that pass one another as they close (scissor action) to cut material such as sheet metal. Saws cut by progressively removing small bits of material as a row of closely aligned "teeth" on a blade are moved against the surface of the material to be cut. Saw blades may be either straight (hand saws) or circular (rotary saws). Drill bits cut by "shaving" their way through material. Hole saws use the rotating motion of a drill to move a type of saw blade through material.

When choosing a cutting tool for forcible entry, several factors must be taken into consideration. First, it must be determined that *cutting* is in fact the most expedient means of gaining entry. Then the point of entry must be chosen. This should be the structural component that yields most easily to forcible entry and, if time allows, that results in the least damage. Once this is determined, the material of which the component is made dictates the type of tool to be used. Other factors, such as accessibility and the availability of a power source, also bear upon the choice of the tool best suited to the situation.

Air Chisel, Air Hammer

The air chisel (Figure 2.30) is a pneumatic device designed to cut sheet metal, but it can be used to cut other materials such as wood and plastic. Powered by compressed air from air bottles or

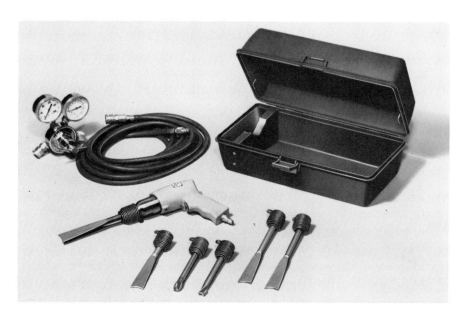

Figure 2.30 Powered by compressed air, the air chisel is especially designed to cut through sheet metal.

compressors, it applies thousands of short-distance impacts per minute to cut metal objects.

WARNING
Although the chisel head seldom causes sparks because of the short distance the bit moves against the objective, do not use the air chisel if there is any possibility that you are working in a flammable atmosphere. The chisel can still cause sparks!

The air chisel is portable and can be put into operation quickly, especially if it is pre-attached to an air cylinder. A 30-minute self-contained breathing apparatus cylinder should last for at least six minutes of continuous air chisel operation. Larger compressor units capable of supplying higher pressure and a continuous supply of air can be installed on a fire apparatus. These units are usually powered by the apparatus drive train through a power takeoff (PTO) device. Systems of this type have a permanently mounted reel equipped with varying lengths of air hose. With this type of system a wider range of air-powered tools can be operated, thus providing the means to cut heavy barriers and to perform a number of tasks that may require large amounts of air. Some compressor units have sufficient capacity to supply air pumps and jackhammers, such as the type used in construction. A compressor unit should *not*, however, be used as an air supply for self-contained breathing apparatus.

Most chisels come in kits that contain a number of bits for performing specialized functions. A long, flat chisel is used for general cutting; a short, flat chisel may be best for situations in which space is limited. A medium-length, panel-cutting chisel, designed with an angled shaft to keep the hands clear of jagged edges, can move quickly through sheet metal. Other attachments can be used for pinching or hammering.

Air chisels are especially well-suited for entry work that involves metal doors, such as warehouse and garage roll-up doors, or metal-clad fire doors. They may also be used for penetrating block walls and for breaking window bar anchors.

The air hammer, or jackhammer (Figure 2.31), is used for much heavier applications, especially masonry construction. Powered by compressed air, the air hammer is generally driven by a large capacity compressor. Smaller gasoline-powered units are also available and are especially useful in areas in which large compressor units cannot easily function. An air hammer is not usually carried on fire apparatus because of its size. If the

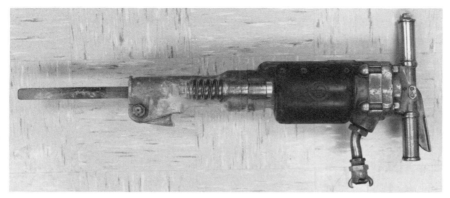

Figure 2.31 The air hammer (jackhammer) is a heavy-duty tool for breaking through masonry walls.

need for an air hammer is anticipated, prior arrangements should be made through other agencies, such as public works departments or rental stores, for access on a 24-hour, short-notice basis.

Axe

Two types of axes, the flat head and the pick head, are perhaps the easiest cutting tools to operate, and most entry crews depend upon them as basic initial attack tools. Quick attack hose evolutions should incorporate an axe as part of the equipment required for successful completion.

The blade of an axe, of course, works most efficiently when it is well sharpened. Sharpening can be done with a cross-mill bastard file and a whetstone (Figure 2.32). Blades should not be sharpened to a keen or "razor-sharp" edge because a keen edge tends to chip. Sharpening an axe at the correct angle to leave sufficient edge density helps prevent chipping (Figure 2.33). Because it is difficult to maintain the proper sharpening angle along the entire width of the blade, power grinders should not be used to sharpen

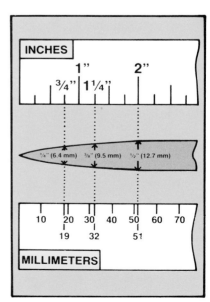

Figure 2.33 Axe blade dimensions after sharpening. To prevent the blade from chipping with normal use, avoid grinding the blade to a keen edge.

Figure 2.32 The cross-mill bastard file and whetstone are the best tools for sharpening an axe.

Figure 2.34 DO NOT USE A POWER GRINDER TO SHARPEN AN AXE. *Courtesy of Bruce Silvera, Chico (CA) Fire Dept.*

axes (Figure 2.34). A blade held against a grinding wheel for more than a few seconds can also heat very quickly. An overheated blade loses its hardness, or temper, and dulls quickly. Grinders also tend to take off more metal than necessary, thus shortening the life of the axe. Some departments use a drill-mounted abrasive disk with a medium grit to rough-sharpen extremely dull axes, then finish sharpening with a file or whetstone. In this case, the motion of the disk across the blade edge is more controlled than it is with a grinder, and a skilled operator can safely sharpen an axe with less chance of removing too much metal.

Wooden axe handles should be maintained by periodic sanding if they become rough or burred, then oiled with a light coat of linseed oil. Handles should be checked frequently for cracks, especially if they are used for prying operations. Prying causes lateral stresses to the grain and can produce hairline cracks that seriously weaken the handle. Fiber glass handles, popular for their resistance to breaking, require little maintenance beyond general cleaning and inspection for chipping. The most vulnerable part of any axe handle is the section adjacent to the blade, called the "shoulder." The shoulder can be wrapped with several layers of inner tube rubber or some other material to protect it when it strikes a hard or jagged surface (Figure 2.35).

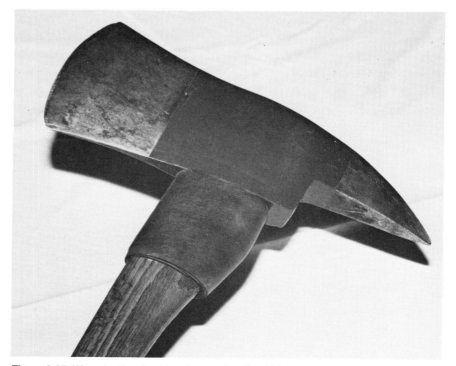

Figure 2.35 Wrapping the shoulder of an axe handle with inner tube rubber protects the wood and extends the life of the handle.

Auger, Drill, and Hole Saw

Augers, drills, and hole saws are used to bore holes into structural materials and locking devices. They may be used to provide a pilot hole for a saw cut (as when a reciprocating saw is used), or for drilling out lock cylinders or other lock components.

Drill bit and augers create relatively small holes by "shaving" small amounts of material when compressed against the surface to be drilled (Figure 2.36). Hole saws (Figure 2.37), in comparison, have sawlike, circular bits for cutting larger holes in doors and walls. Both devices may be either manually or electrically powered.

Another less common type of drill is water-powered. When it is attached to a fire hose, the drill uses water power to prevent sparks. While the drill tends to be awkward to handle, it is useful in situations that require work in hazardous atmospheres.

Manual brace-and-bit drills (Figure 2.38) are completely portable and may be required in situations that are inappropriate for power tools. This tool requires space sufficient to rotate the handle, as well as for the operator to apply force to the head (operator's weight, in some cases), the combined effect of which drives the bit into the material. Hand drills also work well in flammable atmospheres because they are virtually sparkproof.

Figure 2.36 A drill bit creates a hole by "shaving" material as the bit is compressed against the surface.

Figure 2.37 A hole saw works by "sawing" in a circular motion.

Figure 2.38 The brace-and-bit is useful when there is no power source for an electric drill.

Bolt Cutters

Bolt cutters (Figure 2.39) are excellent tools for cutting padlocks, hasps, chains, bolts, and similar materials. Larger models can cut window bars and heavier metal security devices. Most bolt cutters, however, cannot penetrate case-hardened metal, such as that used in expensive padlocks, since the padlock shackle is much harder than the metal in the cutter jaws. A case-hardened padlock can ruin a set of cutter jaws.

Bolt cutters can also be used for cutting wire and cable. Stranded wire and cable is often made of copper or aluminum and, although either may be of a wide diameter, offers little resistance to a sharp pair of cutter jaws.

WARNING
Never cut energized electrical wires or cables with bolt cutters.

Figure 2.39 Bolt cutters are excellent for cutting chains, bars, cables, and other materials that are *not* case-hardened.

Figure 2.40 The oxyacetylene cutting torch is especially useful for cutting heavy metal doors, windows, and gates. *Courtesy of Mike Ridley, Chico (CA) Fire Dept.*

Cutting Torch

The oxyacetylene cutting torch (Figure 2.40) cuts by burning. It is especially useful for penetrating heavy metal doors, windows, and gates that resist more conventional methods. The torch preheats metal to its ignition temperature, then burns a path in the metal with an extremely hot cone of flame.

Naturally, the cutting torch is a specialized tool, and it is not without its hazards. Because it operates with a highly flammable gas and generates a flame, it should be used with extreme caution. Do *not* use a cutting torch if there is any reason to believe that the atmosphere in which forcible entry takes place is flammable. This includes a fire-generated atmosphere that contains combustible gases, as in a flashover situation. As a precautionary measure, it is advisable to have charged hoselines in place before beginning cutting torch operations. Cutting torch operators should be trained to a level of proficiency that ensures the safe and efficient use of the tool in all situations. This requires periodic practice, and anyone who uses the tool should train periodically in exercises that present a variety of cutting problems.

Another hazard associated with the cutting torch is the storage of oxygen and acetylene. Always keep oxygen and acetylene cylinders in an upright position, whether they are in use or in storage. Acetylene is an unstable gas that is pressure- and shock-sensitive. Acetylene storage cylinders, however, are designed to keep the gas stable and safe to use. The cylinders contain a porous filler of calcium silicate, which prevents accumulations of free acetylene within the cylinder space (Figure 2.41). They also contain liquid acetone to dissolve the acetylene, which is stored in liquid form. When an acetylene cylinder valve is opened, the gas leaves the mixture as it travels through the torch assembly hoseline.

WARNING
Keep acetylene cylinders in an upright position to prevent the acetone, a flammable liquid, from flowing through the cylinder valve and pooling in the work area.

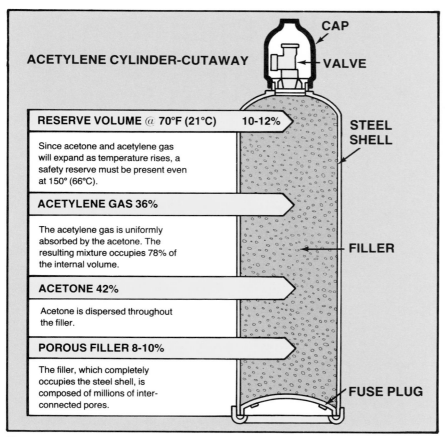

Figure 2.41 The acetylene cylinder has a porous filler of calcium silicate to prevent accumulations of free acetylene within the bottle space. Liquid acetone, a dissolving agent for the acetylene gas, is dispersed throughout the filler. *Courtesy of Victor Equipment Company, Denton, TX.*

The cutting torch generates an extremely hot flame: for preheating metal, the flame temperature in air is approximately 4,200°F (2 316°C); when pure oxygen is added through the torch assembly handle, a flame of over 5,700°F (3 149°C) is created — hot enough to burn through iron and steel with ease.

CUTTING TORCH SAFETY RULES

Observe the following safety rules when using oxyacetylene cutting equipment:

- Store and use acetylene cylinders in an upright position to prevent loss of acetone. Remember, when an acetylene cylinder is empty of acetylene, it is always full of acetone. *Never place empty cylinders on their sides.*

- Handle cylinders carefully to prevent damage to the cylinder and to the filler. A dent in the cylinder indicates that the filler may be damaged. If the filler is damaged, voids are created where free acetylene can pool and decompose — a dangerous condition. Dropping a cylinder may also cause fuse plugs to develop leaks — again, a dangerous condition. Mark and return dented cylinders to the supplier.

- Avoid exposing cylinders to excessive heat. This means that an ambient air temperature exceeding 130°F (54°C) is undesirable for storage or use of acetylene cylinders.

- Avoid placing cylinders on wet or damp surfaces. Cylinders rust at the bottom as protective paint is worn away.

- Store acetylene cylinders in an area separate from oxygen cylinders and other oxidizing gas cylinders. Separate full acetylene cylinders from empty or partially full cylinders. Storage areas should be designed to prevent cylinders from falling when they are bumped (Figure 2.42).

- Perform a soap test (applying a solution of soap and water on fittings) to detect leaks whenever making regulator, torch, hose, and cylinder connections. Slow leaks in confined areas permit acetylene to quickly accumulate in concentrations above the lower flammability limit. Acetylene has a wide flammability range: 2.5% to 81.0% by volume in air. Remove leaking cylinders to an open area immediately. Do not attempt to stop a fuse plug leak.

- Open acetylene cylinder valves no more than ¾ of a turn. Wrenches should not be used on cylinders that have handwheel valves. If the valve is faulty, do not force it. Take the cylinder out of service immediately and return it to the supplier for service.

Figure 2.42 Acetylene cylinder storage areas should be designed to prevent bottles from falling if they are accidentally bumped or pushed.

- Do not use acetylene at pressures greater than 15 psig (103 kPa). Acetylene decomposes rapidly at high pressures and may explode as decomposition occurs.

- Do not exceed the withdrawal rate of 1/7 of the cylinder capacity per hour.

- Keep valves closed when cylinders are not in use and when they are empty. After the valves are closed, bleed off the pressure in the regulator and in the torch assembly. Keep unconnected cylinders capped, whether they are full or empty, to prevent damage to fittings.

Shears

A shearing tool cuts by slicing through material with a scissoring action as the jaw surfaces pass by one another. Shears are especially adaptable for cutting sheet metal and tubing, but they can also cut small rods and bars. Hydraulic shears powered by a gas-powered or electric-powered hydraulic pump are the most commonly used in forcible entry work (Figure 2.43 on next page). Developed for use in rescue and extrication work, (as were their counterpart, spreaders) shears offer a powerful means of penetrating metal doors and window frames.

Like bolt cutters, however, shears have limitations. Case-hardened steel padlocks, for instance, will ruin a set of shears.

Figure 2.43 Hydraulic shears offer a powerful means of cutting during forcible entry. *Courtesy of H.K. Porter Inc.*

They are, however, capable of slicing through the flat metal of a door hasp with little effort.

Forces of up to 15,000 psi (103 425 kPa) can be generated through the blades of hydraulic shears. Some models are designed with special recesses in the jaw surface (Figure 2.44) that are useful for cutting small metal stock or cable.

From a safety standpoint, it should be noted that shears tend to cut in a manner that greatly reduces the hazard of flying material. The material is sliced apart so that the sections gradually drop away from the shears as the cut progresses.

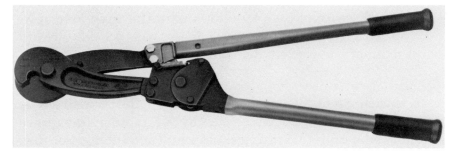

Figure 2.44 Some shears are specifically designed for cutting stranded cable. *Courtesy of H.K. Porter Inc.*

Saws

There are many types of saws, each designed for a specific function. There are two major types of saws: handsaws, which are powered manually, and power saws, which are powered by gas or electric motors.

HANDSAWS

The great advantage of handsaws is their portability. Although they do not cut with great speed, handsaws can be readily used in places where the use of power equipment is not feasible. Handsaws should be chosen according to the type of material to be cut. While this seems an elementary rule for tool selection, it is surprising how often it is overlooked in forcible entry work. Handsaws vary in design and each is intended for cutting a specific type of material at a given rate of speed. Saws designed to produce a smooth, precise cut have many small teeth. Saws intended for less precise cutting have fewer, larger teeth, and cut with greater speed than saws with smaller, more numerous teeth.

Carpenter saws and keyhole saws (Figure 2.45) are used for cutting wood, while hacksaws (Figure 2.46) are used for cutting metal. Most handsaws cut in one direction only. Some hacksaws, however, accept two blades mounted in opposite directions to speed the cutting rate by removing metal in both stroke directions.

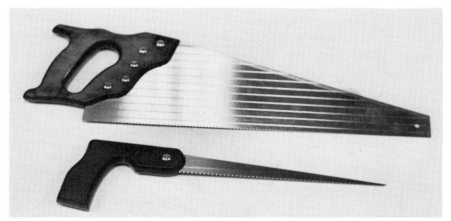

Figure 2.45 The carpenter saw (top) and the keyhole saw (bottom) are handsaws designed for cutting wood.

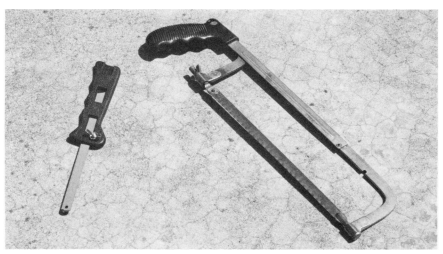

Figure 2.46 Hacksaws are designed to cut metal. The saw on the right has blades mounted in opposing directions so that it cuts with each forward and backward stroke.

POWER SAWS

Power saws are also available in various designs, depending upon the purpose for which they are intended. It is important that the operator know the limitations of each type of power saw. When a saw (or any tool, for that matter) is pushed beyond the limits of its design and purpose, two things may occur: (1) tool failure (including breakage) and/or (2) injury to the operator.

A circular saw (Figure 2.47) is well-suited for cutting a variety of materials because it can be equipped with different blades, each designed for a specific material. As with handsaws, a large-toothed blade produces a faster, less precise cut than a fine-toothed blade. Carbide-tipped blades are superior to standard blades because they are less prone to dulling with heavy use.

A reciprocating saw (Figure 2.48) may be required when overhead work must be done in an area where space is tight. This saw has a short, straight blade that moves forward and backward, with an action similar to that of a handsaw.

Chain saws (Figure 2.49) are gaining in popularity for forcible entry and ventilation work. The best type of saw is one that is powerful enough to penetrate dense material, yet lightweight enough to be easily handled in awkward positions. Chain saws equipped with carbide-tipped chains are capable of penetrating a large variety of structural materials, including light sheet metal. Although carbide-tipped chains cost almost 4 times as much as standard chains, they last 12 times as long.

Figure 2.47 Equipped with the proper blade, a circular saw can cut through almost any kind of material: wood, metal, plastic, or masonry. *Courtesy of Bob Ramirez, Los Angeles (CA) Fire Dept.*

Figure 2.48 A reciprocating saw works well in tight places. *Courtesy of Milwaukee Tool Company.*

Figure 2.49 A chain saw equipped with a carbide-tipped chain has become one of the most popular tools for forcible entry and ventilation.

POWER SAW SAFETY RULES

Following a few simple safety rules when using power saws will prevent most typical accidents:

- Match the saw to the task and the material to be cut. Never push a saw beyond its design limitations.

- Wear proper protective equipment, including gloves and eye protection, when operating a saw (Figure 2.50 on next page).

- Have hoselines in place when forcing entry into an area where fire is suspected. Hoselines are also essential when cutting materials that generate sparks.

- Avoid the use of all saws when working in a flammable atmosphere or near flammable liquids.

- Keep unprotected and unessential people out of the work area.

Figure 2.50 ALWAYS WEAR PROPER PROTECTIVE EQUIPMENT, INCLUDING GLOVES AND EYE PROTECTION, WHEN YOU USE A SAW.

- Follow manufacturer guidelines for proper saw operation.
- Keep blades and chains well sharpened. A dull saw is far more likely to cause an accident than a sharp one.

STRIKING AND BATTERING TOOLS

While forcible entry work requires the application of tools and skills to gain entry quickly and with as little damage as possible, finesse is insignificant when lives are at stake. In such cases, battering and striking tools can penetrate the most stubborn barriers. A number of such tools should be maintained on fire apparatus for forcible entry work.

Striking tools can be used in combination with other tools to facilitate quick entry. Some lock entry tools must be struck so that the sharp edges of their blades "bite" into lock cylinders. Prying tools may require a sharp blow with a striking tool to penetrate a door or window casement and thus provide better leverage.

Striking and battering tools generate tremendous force through the impact part of the tool. This force is the result of (1) the weight (mass) of the tool, and (2) the velocity with which the tool strikes its objective. Force is directly proportional to a combination of both factors: when either weight or velocity is increased, force also increases (Figure 2.51). It follows, therefore, that the best striking and battering tools maximize both factors. Striking tools should be as heavy as can be safely and effectively swung by the operator. If striking tools fail to accomplish the task, it is usually because the operator underestimates the force required to gain entry and chooses a tool that is too light for the job.

There are a number of hazards inherent in striking objects with sudden force. Each blow from a striking tool makes the failure of the object more certain. When an object yields to a striking tool, some portion of the object usually breaks away with a force equal to the force of the final blow. Any tool that must be swung to generate force is, of course, a danger to anyone who is standing in the area. Following a few common-sense safety rules reduces the hazard to those who must be in the working area:

- Provide generous clearance whenever operating a striking or battering tool. Anyone not involved in using the tool should stand well away from the operator and the impact area.
- Wear adequate protective equipment, including eye protection and gloves.
- Maintain all tools to eliminate the danger of broken handles, loose heads, and any condition that contributes to tool breakage or failure under impact.

Figure 2.51 If swung at the same velocity, a 12-pound (5.4 kg) sledge delivers twice the force of a 6-pound (2.7 kg) sledge.

- Maintain full control of the tool when swinging. Overswinging can result in loss of control, which in turn makes the tool ineffective. Choose a tool that is heavy enough to do the job, thereby eliminating the tendency to overswing.

Battering Ram

One of the most effective and safest tools for breaching walls and heavy doors is a battering ram. The battering ram is equipped with handles to allow several people to swing the tool with tremendous force against the objective. The tool is available in varying sizes, depending upon the number of people needed for a breaching operation. A common fire service ram has both a ramming end and a penetrating, or cutting, end (Figure 2.52). A

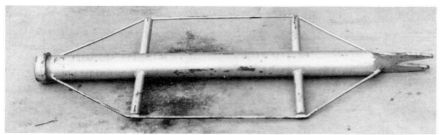

Figure 2.52 A common fire service battering ram has a rounded ramming end and a pointed penetrating end.

police service ram developed for rapid entry through doors during raids is particularly useful for fire service applications. It is used by one or two persons to generate tremendous force so that a door can be penetrated with one or two blows. Its compact shape (Figure 2.53) makes it well suited for working within limited spaces. Larger rams are designed to penetrate heavier materials, such as masonry walls or metal doors.

Figure 2.53 These police service battering rams were designed for rapid entry. Note the compact shape, intended for use by one or two persons. The rams are relatively heavy for their size — an important feature for breaking in a door with one or two blows. *Courtesy of Detroit Police Department, Narcotics Division.*

Hammer and Sledge

Hammers and sledges are available in many sizes and weights (Figure 2.54). For forcible entry work, a short-handled hammer should weigh no less than 6 pounds (2.7 kg). Long-handled sledges come in sizes up to 20 pounds (9.1 kg). When selecting the proper hammer or sledge for the job, it is better to overestimate than to underestimate the size needed. It takes fewer blows with a heavy hammer than with a light one to accomplish a task, and when fast entry is important, the heavier tool is good insurance that the operation will be quickly accomplished. As with any striking tool, hammers and sledges should be swung with short, well-placed strokes, allowing the weight of the tool head to accomplish the task.

The flat-head axe has long been used as a striking tool, primarily because it is usually convenient when a striking task must be done. Because the axe can be used for both cutting and striking, it is considered a multi-use tool. While striking work with a flat-head axe is acceptable in some applications, this is not the use for which the tool was designed. With repeated use for

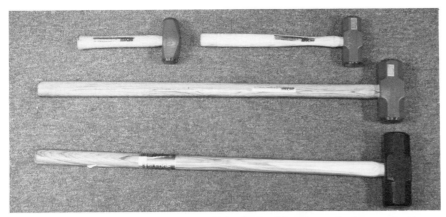

Figure 2.54 When selecting a hammer or sledge for forcible entry work, choose a tool heavy enough to do the job with a minimum number of blows.

striking, it could be damaged. When repeated striking is necessary, it is preferable to use the tool designed for such use — the sledge.

Hammer-Head Pick

The hammer-head pick (Figure 2.55) is an excellent tool for penetrating masonry walls, and in some instances, it can be used as a lock-breaker for padlocks. The hammer-head design allows the pick to be struck with a hammer or sledge to allow more control of the tool. Eye protection is absolutely essential when using a pick, since chips fly from the striking surface with bullet-like speed.

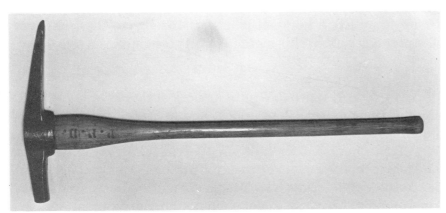

Figure 2.55 A hammer-head pick is used for penetrating masonry walls and for breaking padlocks. *Courtesy of Bob Esposito.*

LOCK-ENTRY TOOLS

Lock entry is becoming one of the most popular methods of forcible entry not only because it accomplishes entry with little damage, but also because it is a surprisingly time-efficient means of entering a locked building. Buildings that are well protected against illegal entry sometimes yield easily to forcible entry through the lock. There are a number of lock-entry tools available that make the job relatively simple. Many of these tools have been designed by firefighters.

Dent Puller

Some locks can be penetrated by removing the lock plug (the part of the lock that receives the key) from the lock cylinder. The dent puller (Figure 2.56), a tool designed for auto body repair, can be used for this purpose. This tool has a small screw end that is twisted into a lock keyhole. Naturally, this tool works best on locks that are made of metals softer than the screw end of the tool.

Figure 2.56 The dent puller, ordinarily an auto body repair tool, can be used to pull key plugs out of lock cylinders.

Figure 2.57 The K-tool is designed for removing lock cylinders, but must be used with another tool, such as the Halligan bar. A key tool is designed to manipulate the interior locking mechanism after a lock cylinder is removed. *Courtesy of Bob Rose, Chico (CA) Fire Dept.*

K-Tool

Another method of lock entry involves pulling the lock cylinder out of the locking mechanism. Designed with a K-shaped, case-hardened blade, the K-tool (Figure 2.57) works by cutting into the outside edges of a lock cylinder, and "biting" into the softer metal of the cylinder. It may then be pried out with a Halligan tool, which provides maximum leverage force. After the lock cylinder is pulled, a key tool is used to disengage the lock latch. Chapter 6 describes this method in greater detail.

Key Tool

Key tools (Figure 2.57) are devices designed to manipulate the interior mechanism of a lock, and to thereby disengage the latching component. As experience is gained in manipulating locking mechanisms, forcible entry personnel may wish to design and fabricate their own key tools for work on locks that are peculiar to their own locales. Such tools are easy to make from materials found in fire department shops and in hardware stores.

O-tool; A-tool

Both the O-tool (an abbreviation for "officer's tool") and A-tool (named for the "A" shape of its head) are similar in design and work on much the same principle as the K-tool. An advantage that these tools have over the K-tool is that they do not require a separate tool for prying the lock cylinder: they have handles that provide a built-in means of leverage. Once the cylinder is removed, a key tool is used to disengage the lock latch. This tool is illustrated in Figure 2.58.

Figure 2.58 The O-tool (A-tool) is also designed for removing lock cylinders. *Courtesy of Fire Fighter Tool Company.*

Figure 2.59 The shove knife is an excellent tool for "loiding" a door latch.

Shove Knife

Entry through simple locking mechanisms can be gained simply by inserting a narrow tool between the door frame and the door to disengage the latch from the strike in the door frame. The shove knife (Figure 2.59) is especially adapted for this purpose. Shaped somewhat like a putty knife, it has a groove cut into one edge of the blade to make positive contact with the latch so the tool does not slip back out of position. It will not work on doors that have rabbetted frames, however, nor on doors that have deadbolt locks.

Answers on page 261

Complete the following statement with words or phrases that make the statement correct:

1. Forcible entry tools are divided into four groups based on the primary manner in which they are used to force entry. The four groups are as follows:

   A. _____

   B. _____

   C. _____

   D. _____

2. The two major advantages of power tools over hand tools are

   A. _____

   B. _____

3. The amount of force exerted by a striking or battering tool is dependent upon two factors:

   A. _____

   B. _____

Answer each of the following questions in a few words or short phrases:

4. To provide maximum mechanical advantage in prying operations,

   A. where should the fulcrum be located in relation to the objective?

   _____

   B. where should force be applied in relation to the fulcrum?

   _____

5. How can a flat-head axe be modified so that it is primarily a prying device instead of a cutting tool?

   _____

6. When using a cutting tool to attempt forcible entry, what is the most frequent reason that the attempt is unsuccessful?

   _____

7. When using a striking tool to attempt forcible entry, what is the most frequent reason that the attempt is unsuccessful?

   _____

8. Indicate whether each cutting tool listed is *relatively* safe or unsafe, based on the potential for generating a spark, for use in a flammable environment.

	Safe	Unsafe
A. Air chisel	☐	☐
B. Flat-head axe	☐	☐
C. Electrically powered auger	☐	☐
D. Hand drill	☐	☐
E. Bolt cutter	☐	☐
F. Shears	☐	☐
G. Power saw	☐	☐

Determine whether the following statements are true or false. If false, state why:

9. When using an air bag for forcible entry operations, obey the following safety rules:

   A. Position the bag only on or against a solid surface.

   ☐ T   ☐ F   _____

   _____

   B. Work under a load supported solely by the bags only if the load weighs less than 100 pounds (45 kg).

   ☐ T   ☐ F   _____

   _____

   C. Do not use air bags where there are materials with surface temperatures above 220°F (104°C).

   ☐ T   ☐ F   _____

   _____

   D. Stack bags no more than three high.

   ☐ T   ☐ F   _____

   _____

10. When storing oxyacetylene cutting equipment, obey the following safety rules:

    A. Store acetylene cylinders in an upright position.

    ☐ T   ☐ F   _____

    _____

    B. Store all cylinders *only* on a dry surface.

    ☐ T   ☐ F   _____

    _____

    C. Store acetylene cylinders in an area away from oxygen cylinders.

    ☐ T   ☐ F   _____

    _____

D. Keep valves open on empty acetylene cylinders.

☐ T ☐ F _____

11. The blade on a pick-head axe should be kept sharpened to a razor sharp edge if it is used primarily for cutting, rather than prying.

☐ T ☐ F _____

12. A charged hoseline should be placed nearby before cutting torch operations begin.

☐ T ☐ F _____

Select the choice that best completes the sentence or answers the question:

13. For optimum mechanical advantage, the length of the head of a lever should be _____ its maximum thickness.
    A. one-half
    B. the same as
    C. three times
    D. six times

14. The door opener is a
    A. lever used to force inward-swinging doors
    B. bar used to spread door casings
    C. lever used to break padlocks and hasps
    D. bar used to remove door hinges

15. Bolt cutters can be used to cut
    A. chains
    B. case hardened padlocks
    C. energized electrical cables
    D. all of the above

16. The speed and precision of a cut made by a saw depends on
    A. whether it is a hand saw or a power saw
    B. the hardness of metal in the saw blade
    C. whether the material cut is softwood or hardwood
    D. the number and size of the saw teeth

17. Which lock entry tool is used to pull a lock cylinder out of the locking mechanism?
    A. A-Tool   C. O-Tool
    B. K-Tool   D. All of the above

18. Which tool should be used to forcibly open a large, sturdy pad-lock?
   A. Hammer-head pick
   B. Bolt cutter
   C. Lockbreaker
   D. Shears

19. The danger from flying debris is greatest with
   A. prying and spreading tools
   B. cutting and boring tools
   C. striking and battering tools
   D. lock entry tools

Match the tool in the left column with the correct term in the right column:

Type of Tool	Powered By
____ **20.** Electric	A. Air
____ **21.** Hydraulic	B. Air/liquid
____ **22.** Pneumatic	C. Liquid
____ **23.** Pneumo-hydraulic	D. Amps/volts

Tool	Classification
____ **24.** Air hammer	A. Prying and spreading
____ **25.** Claw tool	B. Cutting and boring
____ **26.** Dent puller	C. Striking and battering
____ **27.** Halligan tool	D. Lock-entry
____ **28.** Hammer-head pick	
____ **29.** Jimmy	
____ **30.** K-tool	
____ **31.** Kelly tool	
____ **32.** Oxyacetylene torch	
____ **33.** Saw	
____ **34.** Shears	
____ **35.** Shove knife	
____ **36.** Sledge	
____ **37.** Wrecking bar	

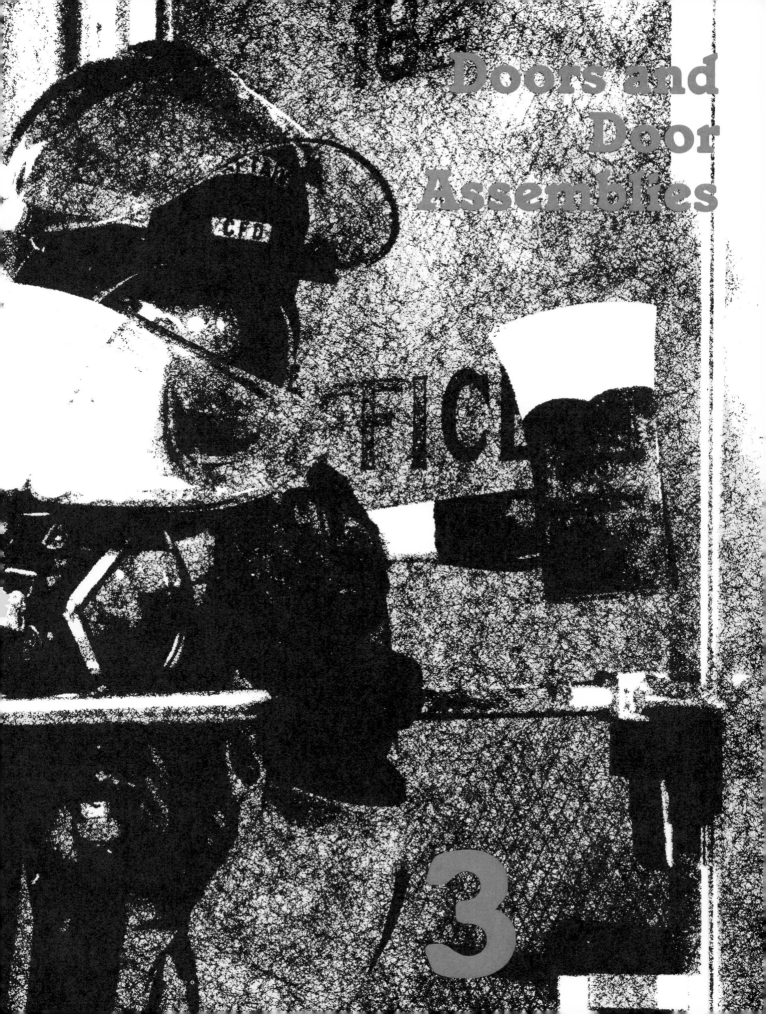

Doors and Door Assemblies

3

NFPA STANDARD 1001
FORCIBLE ENTRY
Fire Fighter II

4-2 Forcible Entry

4-2.1   The fire fighter shall identify materials and construction features of doors, windows, roofs, floors, and vertical barriers and shall define the dangers associated with each in an emergency situation.

Chapter 3
Doors and Door Assemblies

The door has long been the traditional entry point into buildings for fire fighting and other emergency work. Doors, of course, are designed for that purpose — for entrance and for egress. Firefighters know from experience how doors function and therefore choose that component of the building for gaining entrance. When doors are locked, however, a size-up process must take place to determine whether the door should be forced open or if some other point of entry should be chosen. There are basically four options to consider:

- Force a door.
- Force a window.
- Force a lock.
- Breach a wall.

To intelligently evaluate whether forcing a door is the best means of entry, a comprehensive knowledge of how doors are constructed and how door assemblies operate is necessary.

Doors can be broadly categorized into two major groups:

- Residential
- Commercial

Within these groups are a number of different types of doors that can be identified by their components and construction. Doors may also be classified by the manner in which they function. This chapter, therefore, discusses how doors are constructed and how they operate so the forcible entry operator will be able to understand the respective strengths and weaknesses of each type of door. This knowledge is essential to the firefighter who must ask the tactical question, "Is this door the fastest and least damaging means to enter this building?"

RESIDENTIAL AND COMMERCIAL DOORS

A residential door, broadly speaking, is any type of door found in a dwelling, such as a house or an apartment. As a general rule, these doors tend to yield to forcing techniques more readily than doors found in commercial buildings. They are most often made of wood, plastic, glass, or any combination of these materials. Residential door frames are usually constructed of wood (Figure 3.1) and are therefore less resistant to prying than the metal frames of commercial doors.

Figure 3.1 Residential doors are usually made of wood and mounted in wood frames.

Modern manufacturing methods and materials make some residential doors very difficult to force. More importantly, the locking mechanisms for these doors have been improved greatly over the past decade, making residential door entry more difficult than ever before.

Commercial doors are found in all other types of buildings, including warehouses, offices, banks, hospitals, and retail stores. These doors are generally more sturdily constructed for several reasons. Commercial doors receive much more use and abuse than residential doors and must be constructed to last many years under these conditions. They must also be sturdy enough to resist forcing because they are designed to protect property of potentially much higher value than that typically found in residences. Finally, building codes often require that commercial buildings have rated fire doors, most of which are constructed of metal components. These three facts alone indicate that it is usually more difficult to gain entry to a commercial site than to a residence.

Commercial doors are commonly constructed of wood, metal, tempered glass, or combinations of these materials set in wood or metal frames (Figure 3.2). Commercial locks tend to be more sophisticated than residential locks and are therefore more resistant to through-the-lock techniques of forcible entry. Naturally, there are exceptions to every rule. Commercial doors may occasionally be found in residences, and there are certainly many types of commercial occupancies that use residential doors. For the most part, however, firefighters can expect that forcible entry will be more difficult in commercial occupancies. Because each situation is different, firefighters must be familiar with door construction and the methods for forcing different types of doors, and must anticipate special problems through pre-fire planning.

DOOR CONSTRUCTION

Doors can be classified by both design and material. The three most common materials used by door manufacturers are wood, metal, and glass. These materials are often combined to produce a door designed for a particular application. For this reason, doors are grouped first by the predominant material within its structure, then by the design. Table 3.1 contains most typical types of door construction.

Figure 3.2 Commercial doors are often constructed of heavier materials than residential doors and are mounted in wood or metal frames.

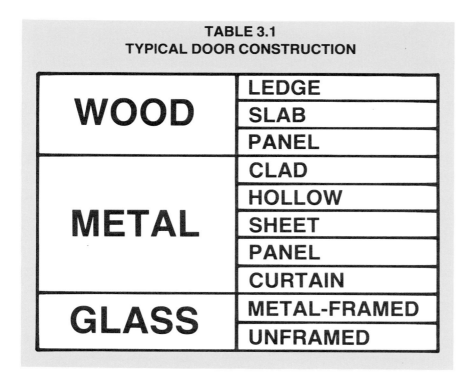

TABLE 3.1 TYPICAL DOOR CONSTRUCTION	
WOOD	LEDGE
	SLAB
	PANEL
METAL	CLAD
	HOLLOW
	SHEET
	PANEL
	CURTAIN
GLASS	METAL-FRAMED
	UNFRAMED

Ledge Doors

One of the oldest door designs is the ledge, or batten, door. Constructed of individual boards joined within a frame, this door is used in such places as warehouses, storerooms, and barns. It can be extremely resistant to forcing operations, depending upon the thickness of the lumber used in the door. Several basic ledge door designs are illustrated in Figure 3.3 on next page.

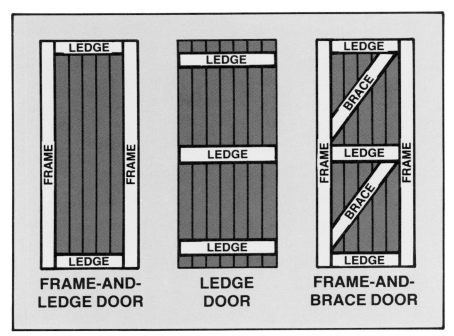

Figure 3.3 Ledge door construction.

Slab Doors

Slab doors appear to be of one solid piece, or slab, of wood. This is rarely the case: few doors are constructed from a single, milled piece of wood. They are constructed of wood components finished to present a smooth, unbroken surface on both sides. There are two basic types of slab doors — hollow-core and solid-core.

The hollow-core door is relatively inexpensive, designed primarily for interior use. Surprisingly, however, many hollow-core doors are used as exterior doors. This makes them much easier to force open with simple tools than solid wood doors. Hollow-core doors are composed of a simple wood frame covered on both sides with a wood veneer (Figure 3.4). The center of the door is hollow, with the exception of a grid of wood or plastic that supports the veneer to help it hold its shape. It is relatively easy to determine whether a door is hollow by simply tapping on its surface. If the door is in fact hollow, this action produces a distinctive hollow sound.

Solid-core doors are formed of laminated blocks of wood or of plied layers of wood and finished with a veneer glued over the core (Figure 3.5). This type of construction increases the dimensional stability of the door, provides thermal insulation, and may boost fire resistance. These doors are commonly found in exterior openings. In some occupancies, such as apartments and motels, where doors along corridors by law must be fire resistant, doors are usually of the solid-core type. These doors, of course, are more difficult to force than hollow-core doors.

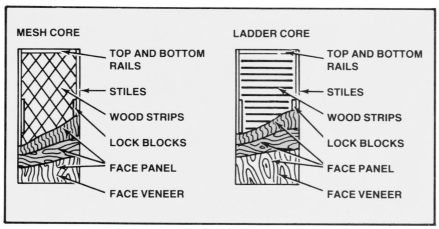

Figure 3.4 Hollow-core doors consist of a frame covered on both sides by a wood veneer.

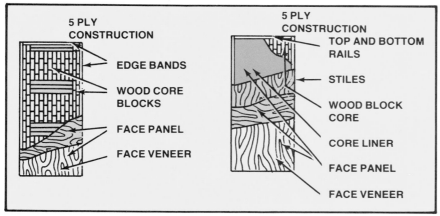

Figure 3.5 Solid-core doors are formed of laminated wood blocks or of plied layers of wood, then covered with a wood veneer.

Wood Panel Doors

Wood panel doors are made of solid wood members inset with panels (Figure 3.6 on next page). The panels may be wood, plastic, glass, or other such materials. Because the panels are relatively thin, they are considered the weakest section of this type of door and should be considered as one of the best places to penetrate during forcible entry.

Metal-Clad Doors

Metal-clad doors, sometimes referred to as "Kalamein doors," are commonly used as fire doors. There are two types: flush design and paneled design. The flush type (Figure 3.7 on next page) consists of metal-covered wood cores; the paneled type is constructed of rails and insulated panels. These doors are more resistant to forcible entry than standard wooden doors because of their metal covering. The metal used to cover the door is 24-gage (0.024 inches [0.61 mm]) or lighter.

NOTE: The insulated panels in this type of door are often composed of metal-covered asbestos. This should be taken into ac-

Figure 3.6 Wood panel doors are made of solid wood members inset with glass, plastic, or wood panels.

Figure 3.7 A flush-type metal-clad door consists of solid wood core covered with 24-gage sheet metal.

count when cutting the panels with a saw. Asbestos is a proven carcinogen and should be avoided if at all possible. Asbestos fibers could be released during the forcible entry operation and inhaled by personnel in the area. If asbestos panels must be cut, self-contained breathing apparatus should be worn by all personnel in the immediate area.

Tin-clad doors are built in much the same way as metal-clad doors, except that the metal is lighter — 30 gage (0.012 inches [0.31 mm]) galvanized steel. The core of these doors is of either two- or three-ply construction.

Hollow Metal Doors

Hollow metal doors are a distinct group of fire doors manufactured of formed 20 gage (0.036 inch [0.91 mm]) or heavier steel. They come in either flush or paneled designs.

Sheet Metal Doors

Sheet metal doors are similar to hollow metal doors, but are made of 22 gage (0.030 inch [0.76 mm]) or lighter steel. They are also available in the flush or paneled designs, but may also be made in a corrugated design (Figure 3.8).

Curtain Doors

Rolling steel doors are made of interlocking steel slats or

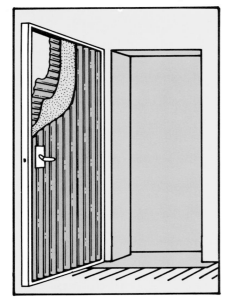

Figure 3.8 Some sheet metal doors are made of corrugated steel; corrugated fire doors have a noncombustible insulating blanket between the metal sheeting.

plates (Figure 3.9). Curtain doors are similar, consisting of interlocking steel blades or of a continuous formed spring steel "curtain" in a steel frame. Curtain doors are often mounted in pairs, one on the outside of the opening and the other on the inside of the opening, to provide maximum fire resistance (Figure 3.10).

Figure 3.9 A rolling steel door is made of interlocking steel plates or slats.

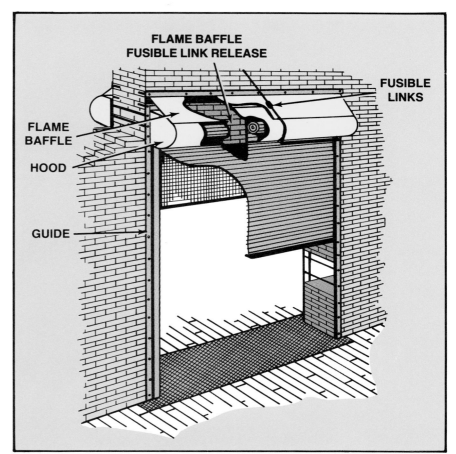

Figure 3.10 Curtain doors, often mounted in pairs, are designed for maximum fire resistance.

Glass Doors

Glass doors are made primarily of glass, usually set in metal frames. Some glass doors have grooves that hold the glass panel into a one-piece tubular frame. The vertical section of a metal frame is called a "stile." Narrow-stiled glass doors contain locks that have latches that swing up or down into position, rather than sliding horizontally from within the frame.

Glass-paneled, tubular aluminum doors set in aluminum frames are becoming common in many commercial occupancies (Figure 3.11 on next page). Lightweight and strong, this type of assembly is highly resistant to prying operations because there is little spring within the frame. These assemblies often contain concealed hinges and locks with a latch throw of over one inch (25 mm), which makes them extremely difficult to force. Through-

Figure 3.11 A well-constructed tubular aluminum frame glass door is difficult to force by prying.

the-lock entry should be seriously considered as the best means to penetrate this type of door.

DOOR GLASS

Glass doors contain one of three types of glass:

- Plate

- Laminated

- Tempered

Each type of glass has different characteristics that affect the way it should be handled when forcible entry requires glass breakage.

Plate glass is the most dangerous type of glass to break because of its tendency to fragment into knife-like shards that can easily cut through clothing and protective equipment. Plate glass is no longer approved for use in glass doors because of the number of serious injuries that have been associated with accidental breakage. Plate glass doors are still in use, however, and any door that cannot be positively identified as containing tempered or laminated glass should be treated as plate glass. Plate glass doors have no identifying marks or characteristics. Tempered glass, however, can be identified by a small decal or stencil in one corner. Laminated glass may also be marked in this way, but this is not always the case.

Laminated glass, also known as "safety glass," is produced by bonding two layers of glass to a thin sheet of vinyl under pressure and heat. When this glass is broken, it tends to remain in its frame with the glass fragments still attached to the vinyl bonding sheet. The glass must then be removed by cautiously pulling it from the frame using some kind of tool (rather than with gloved hands) and wearing protective clothing and eye protection.

Tempered glass is an exceptionally strong glass that is relatively safe to break. Tempering is a treatment process in which glass is heated to just below its melting point, then is cooled rapidly. This places the glass in a state of both tension and compression so that the outer "skin" is much more resistant to breakage than ordinary glass. When broken, however, tempered glass shatters almost explosively because of the internal forces that exist within the pane. It breaks into hundreds of small round fragments that pose less of a cutting hazard than other glass, but remain a serious eye hazard. It is imperative that eye protection be worn during any operation involving tempered glass.

Glass doors are generally expensive to replace; other, less destructive methods of forcible entry, therefore, should be considered before the breaking of glass. Through-the-lock forcible entry is perhaps the best alternative to this method of penetration. Breaking glass is also a hazardous operation in terms of the risk of injury to the eyes and skin. Forcible entry by breaking glass, therefore, should be done only as a last resort. (Chapter 7 contains additional information about glass.)

TYPES OF DOORS

Another way to classify doors is according to the manner in which they operate. This means that for a door to open and close it must move in the fashion dictated by its design. This design, in turn, determines the manner in which the door is attached to the building. A door's design determines the type of frame into which it is set and the hardware that allows it to function as intended. When a door is equipped with hardware and is placed into a frame, it is known as a "door assembly." The lock is the final component in the assembly that must be considered when forcing the door. Lock design and function are discussed in Chapter 5; "through-the-lock" entry methods are discussed in Chapter 6.

Most doors can be classified as one of the following types:

- Swinging
- Overhead
- Sliding
- Revolving
- Folding

Swinging Doors

A common type of door is one that swings to open and close. Swinging doors (sometimes called "hinged doors") have mounting hardware that permits them to pivot on one side of the opening (Figure 3.12). Leaf hinges are the standard type of hardware for these doors, although other devices such as pivot rods can be used. Swinging doors may also be referenced as "right-swinging" or "left-swinging," depending upon which side the hinges are located (Figure 3.13). A right-swinging door has hinges on its right side, as referenced from the outside of the door. It swings in an arc around its hinges to the right side of the opening. Doors may also be called "in-swinging" or "out-swinging," again referenced from the outside. An in-swinging door, commonly found on the exterior of residential structures, must be pushed inward to open (Figure 3.14). Exterior commercial doors are usually out-swinging because life safety codes require this type of door assembly to facilitate rapid egress in a panic situation.

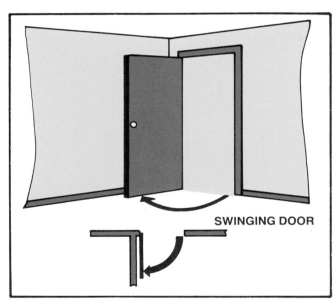

Figure 3.12 The swinging door is designed to pivot to one side of the door frame, usually on hinges.

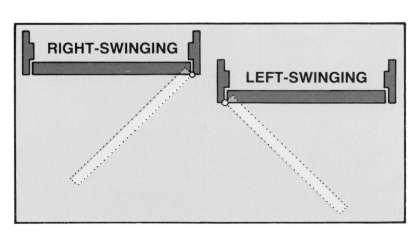

Figure 3.14 An in-swinging door opens toward the interior.

Figure 3.13 As referenced from the outside of the door, a right-swinging door opens to the right side of the frame and a left-swinging door opens to the left side of the frame.

Doors may also be mounted in pairs (Figure 3.15). In this arrangement, one door is sometimes locked in place while the second door swings freely. When both doors are locked, it is usually easier to force the free-swinging door because its locking mechanism is less secure than the anchor door.

Figure 3.15 When swinging doors are mounted in pairs, one door is sometimes locked into place while the second door is allowed to swing free. When both doors are locked, the second door is usually the easier of the two to force open.

DOOR FRAMES FOR SWINGING DOORS

All swinging doors are mounted in frames that permit them to close in a secure manner. Frames for swinging doors have "stops" to prevent the door from swinging through the frame. On the simplest wooden frame, a narrow wooden strip is nailed inside the frame to serve as a stop (Figure 3.16 on next page). Another type of frame has a shoulder piece milled directly into the frame (Figure 3.17 on next page) by a process called "rabbetting." The

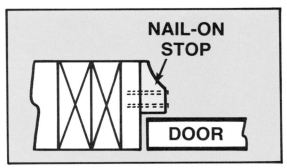

Figure 3.16 The simplest door frames have a door stop that is nailed to the frame.

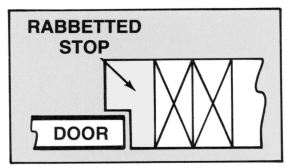

Figure 3.17 Well-constructed wood door frames have a stop that has been rabbetted, or cut, into the frame, making it a part of the frame itself.

Figure 3.18 Metal frames have the stop molded into the frame during the manufacturing process.

frame with a nail-on stop is less resistant to forcible entry than the rabbetted-stop type. Metal frames have the stop formed within the frame as an integral part of the design (Figure 3.18). Metal frames are highly resistant to prying during forcible entry. In-swinging metal doors set in masonry-supported metal frames are extremely strong door assemblies.

Sliding Doors

Sliding doors, while not as common as swinging doors, are often found in commercial occupancies as exterior doors. Residential sliding doors are commonly used as secondary entrance doors, often from porches or patios (Figure 3.19). As a general rule, sliding doors are neither as weathertight nor as soundproof as swinging doors, but many pre-engineered assemblies are very efficient. Sliding doors are installed when it is unsafe or inconvenient to have a door swing into an entry or egress way, or when interior space requirements prevent other types of doors. They may operate manually, electrically, or hydraulically and are usually mounted on overhead tracks (Figure 3.20). Assembly hardware includes a system of rollers, stops, and power equipment as required. Floor-mounted tracks and rollers are less common because of problems associated with accumulations of dirt and debris from the floor. For wide openings, several sliding doors may be mounted on a multitrack system.

Sliding doors are available in several basic designs, based on function and wall requirements. Figure 3.21 illustrates a surface sliding door, which opens over the adjoining wall surface. A bypass sliding door (Figure 3.22) opens over an adjoining door of similar function. This door style is not generally used on exterior openings, but is common on closet or storage room doors. In the open position, the pocket door (Figure 3.23) is hidden within a recess, or "pocket," within the adjoining wall.

Folding Doors

Folding doors are rarely found in exterior openings, although there are some commercial designs available. In general, this type of door is intended only for interior use for many of the same

Figure 3.19 Sliding glass patio doors are secondary entrance doors. New designs are highly resistant to forcible entry.

Figure 3.20 Some large commercial sliding doors operate on overhead tracks.

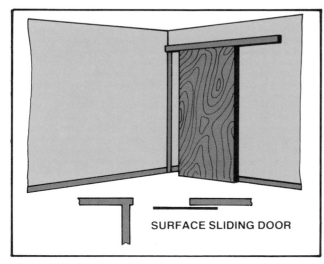

SURFACE SLIDING DOOR

Figure 3.21 A surface sliding door opens over an adjoining wall surface.

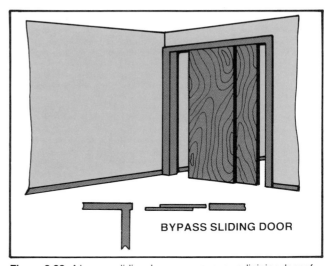

BYPASS SLIDING DOOR

Figure 3.22 A bypass sliding door opens over an adjoining door of similar function.

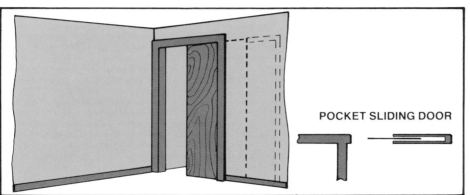

POCKET SLIDING DOOR

Figure 3.23 A pocket sliding door, which retracts into a cavity within the wall, is usually an interior door.

reasons that sliding doors are used. Most folding doors are mounted in overhead tracks that have rollers or glides to support and guide the door during operation.

Multifolding doors, sometimes called "accordion doors," are made of a number of folding sections (Figure 3.24). These doors are used for large openings, such as those between open-walled rooms.

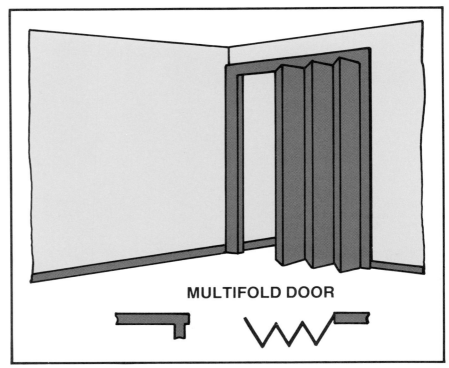

MULTIFOLD DOOR

Figure 3.24 Multifolding, or accordion, doors are constructed of folding sections or panels and are used where space is limited.

Overhead Doors

Overhead doors are designed to cover large openings, such as those in warehouses, stores, and commercial garages. Some types are used as fire doors, that is, they operate automatically in case of fire. Overhead rolling doors are constructed of interlocking metal slats that roll into a storage barrel mounted above the opening (Figure 3.25). The barrel is fitted with a torsion-spring mechanism to counterbalance the weight of the door. Operated by chain, crank, or electric power (Figure 3.26), the rolling door presents a formidable barrier when it is locked. Manually operated models may utilize any number of locking devices for security, ranging from simple slide bars to more sophisticated commercial locks. Electrically operated doors are secured by a lock-down mechanism within the power unit, although the mechanism may be supplemented with locks.

Another type of overhead door is made of hinged panels (Figure 3.27) that move in tracks mounted inside the door opening

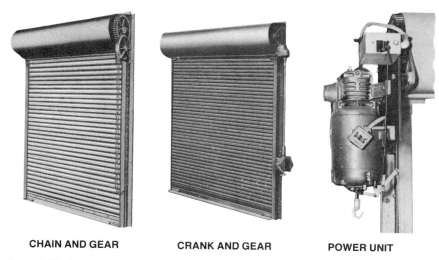

CHAIN AND GEAR   **CRANK AND GEAR**   **POWER UNIT**

Figure 3.26 Overhead doors are operated by chain, crank, or electric power. The power unit usually contains a locking mechanism that prevents the door from being pried open.

Figure 3.25 Overhead rolling doors are made of interlocking metal slats that roll into a storage barrel when the door is raised.

Figure 3.27 Another type of overhead door is made of hinged panels that move on tracks inside the door opening and along the ceiling.

and along the ceiling. This design is common in garage doors, but is also found in commercial occupancies such as warehouses and automobile repair shops. Such doors are either manually operated or may be powered by electric door openers. They may be secured with surface bolts that lock the door to its tracks. Deadbolt locks or even padlocks may also be used, mounted on either side of the door.

Garage doors are often constructed in one piece and are commonly known as "slab" doors. This type of door swings overhead on articulating hardware that incorporates large springs to counterbalance the weight of the door.

Revolving Doors

A revolving door presents a unique problem to firefighters who must make entrance for fire suppression activities: when the door functions normally it does not allow for the movement of hose into the structure. In the event of fire, the door must be disabled so that it does not revolve within its housing.

Revolving doors are constructed of several sections, or wings, that rotate on a central pivot (Figure 3.28) within a semicylindrical housing. They are designed to provide a climatic barrier for the building served, since the opening is essentially closed to the elements no matter what the position of the door sections. This is especially important in high-rise buildings in which a "stack effect" must be maintained. Modern multistory buildings are pressurized with heated or cooled air, depending on the outside temperature; thus, a conventional door allows a large volume of internal air to quickly pass to the outside. Revolving doors, on the other hand, allow relatively little internal air to pass to the outside and thereby decrease energy waste significantly in high-rise buildings.

Revolving door wings are secured into place within the unit by any one of several types of mechanisms. Older models may em-

Figure 3.28 A revolving door has three or four sections, or wings, that rotate within a semicylindrical housing on a central pivot.

ploy simple chain keepers or stretcher bars between the wings
(Figure 3.29), while newer models utilize spring-loaded cam-in-
groove or bullet-detent hardware (Figure 3.30). Most models employ a collapsing mechanism that allows the wings to "book fold"
to one side when the wings are pushed in opposite directions (Figure 3.31). This feature is required to satisfy life safety codes that
provide for rapid exiting from a building. The amount of force required to collapse the wings varies from 50 to 150 pounds (222 N
to 667 N).

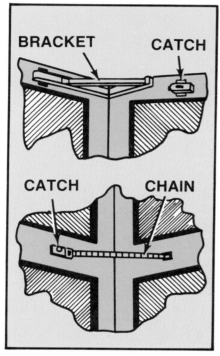

Figure 3.29 Older revolving doors have
stretcher bars (top) or chain keepers (bottom)
between the wings that hold them in place.

Figure 3.30 Modern revolving doors use a cam-in-groove or bullet-detent mechanism to
hold the wings in place. *Courtesy of Ray C. Goad, Goad Engineering, Boonville, IN.*

Revolving doors may be locked with simple slide bolts, mortise locks, or other types of customized locks. Although through-
the-lock entry is preferable to breaking glass within the door
units, due to the extreme high cost of glass replacement, locks
may not be accessible from the outside (Figure 3.32 on next page).
They are sometimes mounted on the inside of two opposing wings
of the door. When this style of lock is employed, the person securing the revolving door leaves the building through another exit.
In this case the best option for forcible entry personnel is to seek
an entry point other than the revolving door.

Fire Doors

Standard fire doors are of the following types:

● Horizontal and vertical sliding

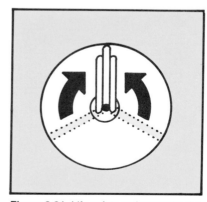

Figure 3.31 Life safety codes require the
wings of revolving doors "book fold" to one
side when two wings are pushed in opposite directions.

Figure 3.32 Revolving door locks are usually only accessible from the inside. *Courtesy of Ray C. Goad, Goad Engineering, Boonville, IN.*

- Single and double swinging
- Overhead rolling

Fire doors may be counterbalanced and can work two ways: (1) self-closing doors automatically close after someone opens and passes through the opening, and (2) automatic-closing doors normally stay open but close when heat activates a closing device, such as a fusible link or electromagnetic catch.

Fire doors that slide horizontally are preferable to other types when floor space is limited. They operate on overhead tracks that are mounted in such a way that when a fusible link releases the door, a counterweight causes the door to move across the opening (Figure 3.33). Vertical sliding fire doors are normally used where horizontal sliding doors or swinging doors cannot be used. Some models utilize telescoping sections that slide into position vertically on side-mounted tracks; the sections are operated by counterweights (Figure 3.34). Both horizontal and vertical fire doors close automatically.

Swinging fire doors are generally used in stairwells (Figure 3.35) and other places where they must be opened and closed frequently in normal service. Swinging doors fit tightly into a rabbetted jamb and thus more efficiently prevent penetration of smoke and fire gases than sliding fire doors.

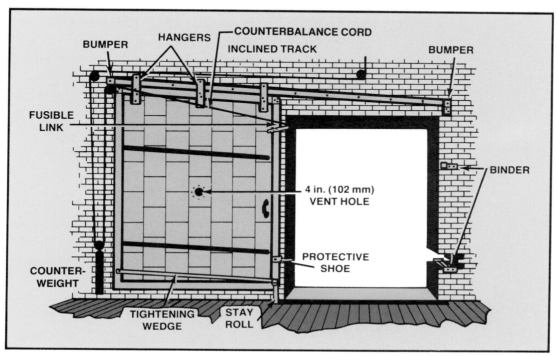

Figure 3.33 A sliding fire door is used where there is not enough space for a swinging fire door.

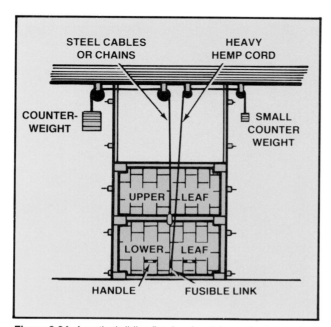

Figure 3.34 A vertical sliding fire door has telescoping leaves that slide on side-mounted tracks.

Figure 3.35 Swinging fire doors are usually found in stairwell openings.

Swinging fire doors are found in both single and double assemblies (Figure 3.36). They are designed to close automatically when triggered by either heat or by an electrical signal. Modern swinging fire doors utilize electromagnetic catches that de-energize to release the doors, which are equipped with automatic closers (Figure 3.37). These mechanisms operate when there is a power interruption or upon receipt of an electrical signal from an automatic alarm such as a heat or smoke detector.

Overhead rolling fire doors can be installed where space restrictions prevent the installation of vertical sliding doors. Counterbalanced fire doors are generally used in freight elevator openings, mounted on the face of the wall inside the shaft. They also close automatically.

Fire doors encountered in the closed position during entry operations should be approached with caution. Although fire doors are rarely locked, they may be the only barrier between firefighters and a hostile environment on the other side of the opening.

Figure 3.36 Double swinging fire doors, common in the corridors of commercial buildings, automatically close when alarm systems are triggered.

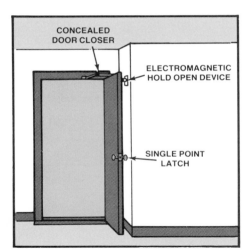

CONCEALED DOOR CLOSER

ELECTROMAGNETIC HOLD OPEN DEVICE

SINGLE POINT LATCH

Figure 3.37 A swinging fire door is equipped with an electromagnetic catch, which works by de-energizing when it receives an alarm signal. This releases the door, causing it to close automatically.

Answer each of the following questions in a few words or short phrases:

Review
Answers on page 262

1. What are the four basic components against which forcible entry is usually attempted?

   A. _____

   _____

   B. _____

   _____

   C. _____

   _____

   D. _____

   _____

2. Doors are categorized into two major groups, commercial and residential. What are the three different ways in which doors are identified within these groups?

   A. _____

   _____

   B. _____

   _____

   C. _____

   _____

3. What are the three most common materials used in the manufacture of doors?

   A. _____

   _____

   B. _____

   _____

   C. _____

   _____

4. Doors may have a hollow-core or a solid-core. Indicate whether each type of door listed is *only* found in the hollow-core (HC) design, *only* in the solid-core (SC) design, or may be found in both (B) designs.

	HC	SC	B
A. Wood, ledge	☐	☐	☐
B. Wood, slab	☐	☐	☐
C. Wood, panel	☐	☐	☐
D. Metal, sheet	☐	☐	☐

5. Doors are classified according to the manner in which they operate. What are these five classifications?

A. _____

B. _____

C. _____

D. _____

E. _____

6. Double swinging doors are often arranged so that one door is locked in place while the other swings freely. Which of the two — the anchor door or the free-swinging door — is easier to force when both are locked in place? Why?

_____

_____

7. What is the unique problem presented by an unlocked revolving door during fire fighting operations?

_____

_____

8. What are the three standard types of fire doors?

A. _____

B. _____

C. _____

Determine whether the following statements are true or false. If false state why:

9. Forcible entry is more difficult on doors of commercial occupancies than on doors of residences.

☐ T   ☐ F   _____

_____

10. The forcible entry method of breaking the glazing out of a glass door should not be considered as a primary technique.

☐ T   ☐ F   _____

_____

Select the choice that best completes the sentence or answers the question:

11. Because the door may be filled with asbestos, self-contained breathing apparatus should be worn if it is necessary to cut open a
    A. metal curtain door
    B. hollow metal door
    C. flush design, metal-clad door
    D. panel design, metal-clad door

12. Which door frame is usually the easiest to force?
    A. Wood frame with a nail-on stop
    B. Wood frame with a rabbetted stop
    C. Metal frame
    D. All are equally resistant to forcing

13. Which type of door is most often used to cover large openings in warehouses, stores, and commercial garages?
    A. Folding
    B. Overhead
    C. Sliding
    D. Swinging

Complete the following statements with words or phrases that make the statements correct:

14. The exterior door in the sketch below is classified as

    _____ and _____.
    (right-swinging, left-swinging)   (in-swinging, out-swinging)

15. The bathroom door in the sketch below is classified as

    _____ and _____.
    (right-swinging, left-swinging)   (in-swinging, out-swinging)

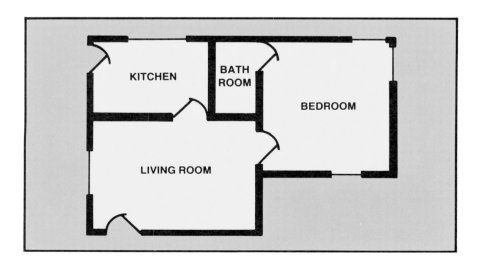

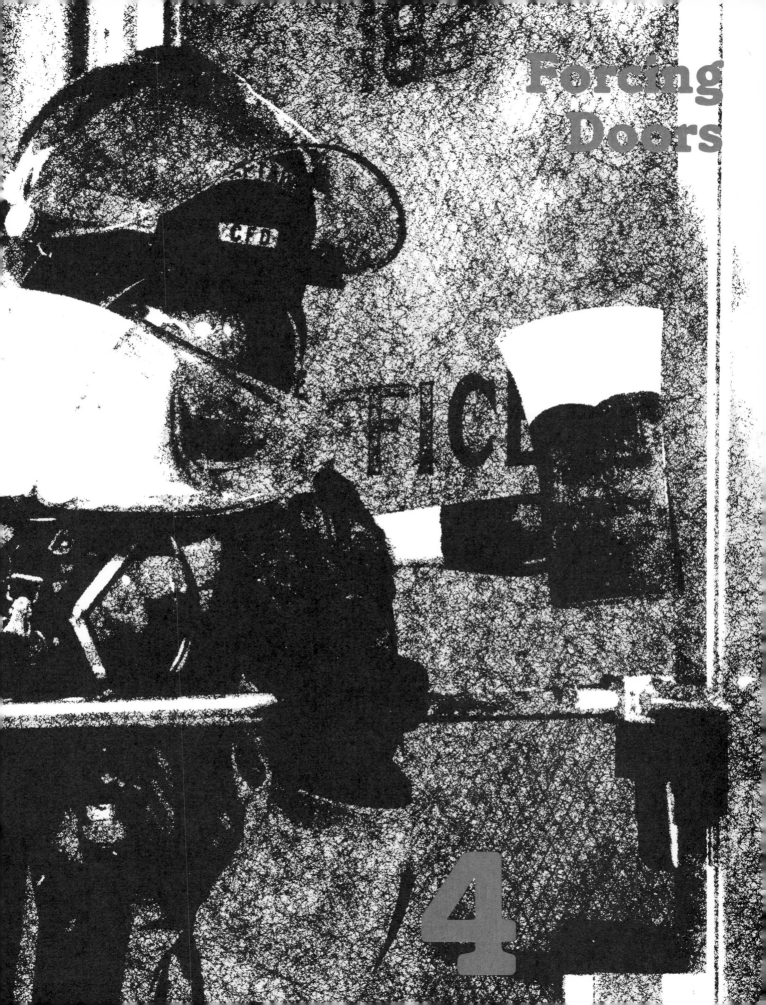

Forcing
Doors

4

NFPA STANDARD 1001
FORCIBLE ENTRY
Fire Fighter I

3-2 Forcible Entry

3-2.1   The fire fighter shall identify and demonstrate the use of each type of manual forcible entry tool.

Fire Fighter II

4-2 Forcible Entry

4-2.1   The fire fighter shall identify materials and construction features of doors, windows, roofs, floors, and vertical barriers and shall define the dangers associated with each in an emergency situation.

4-2.2   The fire fighter shall identify the method and technique of forcible entry through any door, window, ceiling, roof, floor, or vertical barrier.

Chapter 4
Forcing Doors

Once fire fighting personnel have decided that forcing a door is the most effective way to enter a locked building for a given situation, they must decide *which* door to force. The front door, often the most accessible opening upon arrival, may not be the best place to enter a building. The evaluation of a number of factors allows the entry crew to make entry at the most strategic point. Such factors include the type of door and lock assemblies, fire location within the building, and the distance between the entry point and the pumping apparatus.

Modern fire fighting tactics advocate an aggressive *interior* attack whenever possible. The objective of an interior attack is to confine the fire to the immediate area of involvement. To accomplish this, hoselines must be advanced through the interior so that fire is not pushed into an area of noninvolvement. The correct placement of hoselines often requires that entry be made through a door that is not as accessible or as easy to force as other doors.

Forcible entry is a key element in the success of a fire attack operation. If it takes a crew longer to force a door than expected, fire intensity can escalate beyond the capabilities of the attack resources, resulting in the loss of the structure. Successful entry, then, means not only penetrating the opening, but also accomplishing the operation quickly.

Another element of successful door entry is minimizing damage. It can be argued that damage is expected in a situation that requires speedy entry, and that damages can be written off as a justified fire loss. While this viewpoint is shared by many firefighters, it bears some criticism. A similar philosophy regarding the use of water dominated the fire fighting profession not many years ago. Many firefighters practiced the indiscriminate application of excessive volumes of water into fire-involved buildings with little thought for the damage they caused. Today, suppression methods have changed so that minimal amounts of water are used to control and extinguish fire, which in turn minimizes water damage. This change did not occur until firefighters began to accept that the job could be done more *efficiently*.

Efficient forcible entry, then, involves not only speed, but also minimal damage. Minimizing damage does not require that an operation be performed slowly. On the contrary, a method that results in little or no damage is likely to be completed in a very short time, given the proper tools and training. Proper training requires a thorough understanding of door function and design, as well as practice with forcible entry tools in methods that involve prying, cutting, and striking.

Once entry crew members select a door as the entry point to a building, they should perform a simple but effective operation: *attempting to open the door in the normal manner*. There is no reason to assume that a closed door is a locked door. The most basic rule for forcible entry is "TRY BEFORE YOU PRY!" (Figure 4.1). A simple turn of a doorknob may be the only effort required to open a door. (Unlocked wooden doors may require an additional shove because of their tendency to swell during damp or humid weather.)

TRY BEFORE YOU PRY!

Figure 4.1 The most basic rule of forcible entry is . . .

At the "try" stage of entry, make a mental note if the door shows signs of previous forcible entry. If so, it may indicate that illegal entry was made before the fire department's arrival. This information may be of importance to fire investigators after the incident is over. Similarly, an unlocked door should be reported, especially if the door is one that should have been locked, such as the entry to a business after hours. Touch the door lightly with one hand, first at the bottom, then move upward (Figure 4.2). High temperatures indicate that fire may be just beyond the opening. Be prepared with hoselines and full protective equipment. Remember to check for signs of backdraft, such as smoke puffing from door cracks. If there is any doubt about the safety of opening a door in a suspected backdraft situation, *DO NOT ATTEMPT ENTRY*. Notify the incident commander so that vertical ventilation can be performed before entering the building.

Figure 4.2 In a potential fire situation, check the door for heat before forcing it open.

SIZING UP THE DOOR

When determining the best method for forcing a door, several observations should be made:

How is the door constructed?

- Wood, metal, glass
- Solid, hollow, panel

How does the door operate?

- Swinging
- Sliding
- Folding
- Rolling
- Revolving

In what type of frame is the door mounted?

- Wood
- Metal
- Adjacent wall materials

What type of hardware does the assembly contain?

- Hinges
- Pins
- Rollers
- Guides
- Pivots

What types of locking devices are in use?

- Padlocks
- Deadbolt locks
- Bars
- Surface bolts
- Manually or electronically operated

FORCING SWINGING DOORS

The first thing to determine when a swinging door must be forced is the direction in which it opens. To determine if a door opens inward or outward, check the door frame and hinges. If the door opens outward (toward the person seeking entrance), the stop on the frame cannot be seen from the outside and the hinges are usually visible. An in-swinging door has a frame stop on the exterior side of the door and hinges that are not visible from the outside.

When hinges are accessible, one of the first methods to consider for forcing the door consists of simply removing the hinge pins. Any type of tool with a narrow blade (a screwdriver works well) can be used to pry the pin from the barrel of each hinge plate. Starting with the bottom hinge, insert the blade between the plate and the head of the hinge pin (Figure 4.3). This may require the use of a striking tool to force the blade in sufficiently to reach the pin shaft. Lever the pin out or use a striking tool to force the blade and pin upward (Figure 4.4). Continue upward to the next hinge, repeating the process. When all pins are removed, the door can be pulled outward on the hinged side, freeing the lock latch from the door frame (Figure 4.5). Doors that have only two hinges often require the removal of only one pin to release the door from the frame to the point that the latch clears its strike and the door can be swung open.

Figure 4.5 Pry the door loose from the hinge side, and pull the door outward to free the door latch from its strike.

Figure 4.3 To pry a hinge pin, first insert a screwdriver blade between the hinge plate barrel and the head of the hinge pin.

Figure 4.4 If necessary, strike the end of the screwdriver handle with a hammer to move the pin out of the barrel.

Prying is one of the best methods for forcing a swinging door. The size-up of a swinging door is usually made with this method in mind, although other methods should also be considered. Prying tools with large adz surfaces and long handles are best for forcing swinging doors, especially when the door is of solid core or metal construction and is mounted in a metal frame.

The primary objective in prying a swinging door on the lock side is to move the door and frame apart so that the latch is disengaged from the strike. During the size-up process it is important that the entry crew identify the locking device and its latch length. Doors equipped with locks that incorporate latches of up to 3/4 inch (19 mm) in length can be pried with moderate success if the door and frame are flexible enough to be spread this distance. However, if latches of greater length are used in the assembly, prying at the lock side may not be the best way to force the door.

Prying on the hinged side of the door should also be considered, primarily because some hinges are fastened to the frame with relatively short screws. Prying the door at each hinge breaks all points of attachment so that the door can be pulled or pushed free on that side. When using this method, start at the bottom hinge and work upward to the topmost hinge to allow for maximum control of the door.

Prying Out-Swinging Doors

This method works well on less well-constructed door assemblies, but variations may be needed to force sturdier doors. Two tools can be used alternately, one above the lock and the other below the lock:

Step 1: Insert the adz of the prying tool between the door and frame near the latch (Figure 4.6).

Step 2: Work the adz in against the stop. If necessary, drive the tool inward with a sledge or flat-head axe until it reaches the stop (Figure 4.7).

Step 3: Apply leverage to the prying tool handle in a direction away from the door to separate the door from the frame (Figure 4.8). Placing a fulcrum near the adz offers greater mechanical advantage. Insert another adz tool in the opening on the other side of the latch, if necessary, and repeat the prying process until the latch clears the strike.

Step 4: Pull the door open slowly as the latch leaves the strike.

Figure 4.6 STEP 1: Insert the adz of the prying tool between the door and frame near the latch.

Figure 4.7 STEP 2: Work the adz in against the stop. If necessary, drive the tool inward with a sledge or flat-head axe until it reaches the stop.

Figure 4.8 STEP 3: Apply leverage to the pry tool handle in a direction away from the door to separate the door from the frame. If necessary, insert another adz tool in the opening on the opposite side of the latch and repeat the prying process until the latch clears the strike.

Another method of prying an out-swinging door incorporates a different principle — *twisting* a tool to separate the door and frame. Although this can be accomplished with a standard axe, the best tool for the job is a pry-axe, because it is specifically designed for twisting. To force a door with a pry-axe, follow these steps:

Step 1: Insert the head of the pry-axe between the door and frame near the latch. Work the blade in against the stop. If necessary, drive the tool to the stop with a striking tool (Figure 4.9).

Step 2: Apply leverage to the pry-axe by using the auxiliary handle, twisting the axe head to separate the door and frame until the latch clears the strike (Figure 4.10).

Step 3: As the latch leaves the strike, pull the door open slowly.

Figure 4.9 STEP 1: Insert the head of the pry-axe between the door and frame near the latch. Work the blade in against the stop. If necessary, drive the tool to the stop with a striking tool.

Figure 4.10 STEP 2: Apply leverage to the pry-axe with the auxiliary handle and twist the axe head to separate the door and frame until the latch clears the strike.

Prying In-Swinging Doors

Hinged doors that open away from entry personnel are more difficult to force than out-swinging doors because the stop blocks access to the door edge and to the latch. If the frame has a nail-on stop, it is often possible to pry it from the frame to gain access to the door edge and latch for prying. In this case, the prying operation is similar to that for an out-swinging door. Using a sharp-edged adz, follow these steps:

Step 1: Bump the edge of the adz against the stop to break the paint or varnish (Figure 4.11 on next page).

Step 2: Pry the stop away from the frame (Figure 4.12 on next page) and remove it.

Step 3: Force the blade between the door edge and the frame near the latch (Figure 4.13).

Step 4: Apply leverage to the prying tool handle to separate the door from the frame. If necessary, insert another adz tool in the opening on the opposite side of the latch and repeat the prying process until the latch clears the strike.

Step 5: As the latch leaves the strike, push the door open slowly.

Figure 4.11 STEP 1: Bump the edge of the adz against the stop to break the paint or varnish.

Figure 4.12 STEP 2: Pry the stop from the frame and remove it.

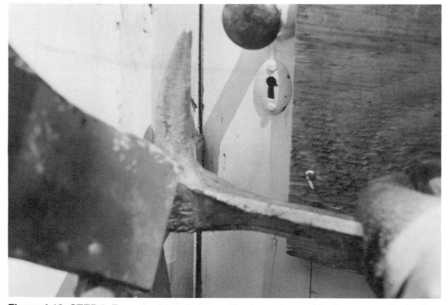

Figure 4.13 STEP 3: Force the blade between the door edge and the frame near the latch.

An in-swinging door set in a frame with a rabbetted stop is difficult to pry because the stop is an integral part of the frame, and it cannot be pried away to allow access to the door edge and latch. In well-fitted doors, such as metal doors set in metal frames, tools cannot easily be inserted between the stop and the door edge. Under these circumstances, two tools must be used to progressively pry apart the door and frame. Tools with sharp-edged adz surfaces work best to penetrate the narrow space between the stop and the door. The steps are as follows:

Step 1: Force the adz blade between the stop and the door near the latch (Figure 4.14). If necessary, use a striking tool to drive the blade in far enough to reach the door frame. If the frame is made of wood, avoid driving the adz blade into the wood.

Step 2: Apply leverage to the tool, using the stop as a fulcrum to move the door away from the frame.

Step 3: Maintaining pressure on the first tool's handle, force the blade of a second tool into the widened opening between the door edge and the frame (Figure 4.15).

Step 4: Apply leverage to the second tool, using the stop as a fulcrum, to move the door away from the frame. Repeat this prying process until the latch clears the strike.

Step 5: As the latch leaves the strike, push the door open slowly.

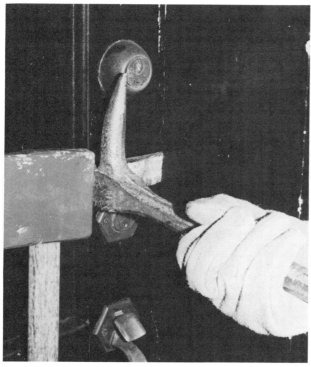

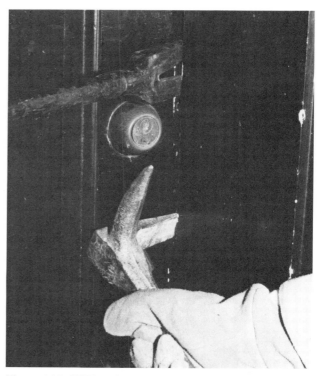

Figure 4.14 STEP 1: Force the adz blade between the stop and the door near the latch. If necessary, use a striking tool to drive the blade in far enough to reach the door frame.

Figure 4.15 STEP 3: Maintaining pressure on the first tool's handle, force the blade of a second tool into the widened opening, between the door edge and the frame.

Prying Double Swinging Doors

Double swinging doors are common in commercial occupancies and are often considered first as the point of entry. Although through-the-lock entry is usually the best method to force a double swinging door, prying may be necessary in some cases. These doors are typically constructed of metal, glass, or some combination of these materials. Unlike single swinging doors, double swinging doors do not latch into the frame. One or both doors may be free-swinging in the unlocked position. Occasionally, one door is fitted with a vertical strip of molding (astragal) that acts as a stop for the companion door.

Locking devices for double doors may vary from a simple bar set in brackets to an elaborate combination of vertical-throw surface bolts or deadbolts. In any case, the objective in prying a double door is similar to a single swinging door — to spring the door far enough to allow the latch to slip out of its strike.

When double doors are secured with only an inside horizontal bar, it may be possible to dislodge the bar by slipping a narrow tool between the door sections and gently lifting or sliding the bar from its brackets. When doors are locked by vertical bolts that slide into strikes that are recessed into the floor and upper frame, the task of forcing is much more difficult. If both doors are secured in this manner, prying is probably not the best means of forcing the door. A cutting operation on one door may be the best way to gain access to the inside bars. If only one door is secured with vertical bars and the companion door locks into this anchor door with a conventional deadbolt lock, prying the latch from the strike often works. In this case, it may be necessary to remove an astragal from the anchor door to allow access to the door edge.

A prying method that involves the use of a hydraulically powered multi-use tool known as the "rabbit tool" is described below. This tool is specifically designed to pry *in-swinging doors mounted in metal frames*, although it has many other applications in forcible entry work.

Step 1: Insert the rabbit tool jaws next to the lock between the stop and the door (Figure 4.16). This can usually be done by driving the jaws in with the palm of the hand; for a tight-fitting door use a striking tool. If there is more than one lock, place the jaws between locks.

Step 2: Place one foot on the leg of the pump while holding the jaws in place with one hand, and grasp the pump handle with the free hand (Figure 4.17).

Step 3: Pump the portable hydraulic pump while maintaining a grip on the back of the jaws (Figure 4.18). Hold the jaws in place throughout the entire operation to prevent them from slipping out of position as they spread. Be prepared for the door to open suddenly as the latch clears its strike.

Figure 4.16 STEP 1: Insert the rabbit tool jaws next to the lock between the stop and the door by driving the jaws in with the palm of the hand; for a tight-fitting door, use a striking tool. If there is more than one lock, place the jaws between locks.

Figure 4.17 STEP 2: Place one foot on the leg of the pump while holding the jaws in place with one hand; grasp the pump handle with the free hand.

Figure 4.18 STEP 3: Pump the portable hydraulic pump while maintaining a grip on the back of the jaws. Hold the jaws in place throughout the entire operation to prevent them from slipping out of position as they spread.

Another method of opening an in-swinging door uses a tool that has been around for many years. The Detroit door opener was developed for such forcible entry operations as pushing open doors, spreading door frames, and breaking padlocks and hasps. Figure 4.19 lists the parts of the door opener. The steps for operating the Detroit door opener are as follows:

Step 1: With the door opener in the closed, vertical position, stand it near the door handle or lock with the fulcrum resting on the floor. Place a finger on the tool at the height of the door handle or lock (Figure 4.20).

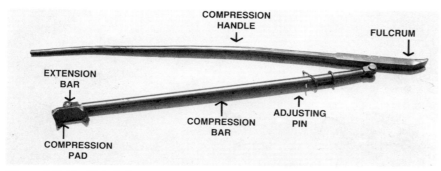

Figure 4.19 The Detroit door opener.

Figure 4.20 STEP 1: With the door opener in the closed, vertical position, stand it near the door handle or lock with the fulcrum resting on the floor. Place a finger on the tool at the height of the door handle or lock.

Step 2: Keeping the finger in place on the tool, lay the tool on the floor perpendicular to the door with the fulcrum against the door. Note the point on the floor upon which the marking finger falls (Figure 4.21).

Step 3: Place the fulcrum on the indicated point and position the compression pad directly under the door handle or lock (Figure 4.22).

NOTE: If the floor surface is slick, as with smooth-finished concrete, it may be necessary to anchor the fulcrum by placing a foot against it or, in extreme situations, by chipping an indentation into the floor surface at the fulcrum spot.

NOTE: If necessary, remove the adjusting pin and slide the extension arm in or out to the proper length. Reinsert the adjusting pin so that with the compression pad in position, the compression handle is at the proper angle to apply pressure. The compression bar and the compression handle should be at the same angle in relation to the floor.

Step 4: Apply increasing force to the compression handle (Figure 4.23) until the door breaks free.

Figure 4.21 STEP 2: Keeping the finger in place on the tool, lay the tool on the floor perpendicular to the door with the fulcrum against the door. Note the point on the floor upon which the marking finger falls.

Figure 4.22 STEP 3: Place the fulcrum on the indicated point and position the compression pad directly under the door handle or lock.

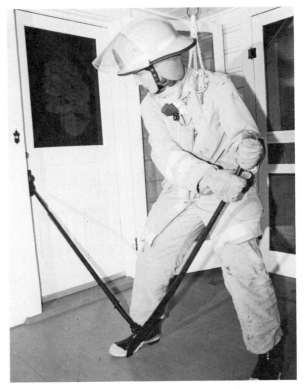

Figure 4.23 STEP 4: Apply increasing force to the compression handle until the door breaks free.

Cutting Swinging Doors

Cutting a door is an effective way to gain entrance quickly. While it may seem that cutting is a very destructive method of forcible entry, this is not necessarily so. Doors with panels offer a means to cut an access hole with little total damage. The hole need only be large enough to allow personnel to reach inside to the locking mechanism and unlock the door. The damaged panel can later be replaced at a nominal cost.

When cutting to gain access to securing devices, use a power saw to cut a triangle in the door near the lock (Figure 4.24). Beware of the hazards associated with starting a cut on a flat vertical surface. Every type of saw reacts differently to this angle of cutting. Wear full protective clothing, including gloves and eye protection.

An alternative to the saw is the hole saw, which can create a fist-sized hole. Smaller drills can also be used to drill a circular pattern of holes in the door that are close enough to one another so that the section of door inside the pattern can be struck with a tool and removed (Figure 4.25).

Another cutting method results in total destruction of the door, but is one of the fastest ways to gain entrance. This method involves cutting the door vertically from top to bottom. The cut divides the door into two sections, one of which can be removed from the frame. Make the vertical cut in the center of the door, starting at the top and cutting downward (Figure 4.26).

Figure 4.24 Use a power saw to cut a triangle in the door near the lock. After the hole is cut, reach inside to unlock the door.

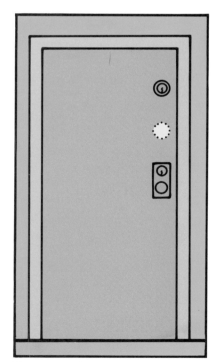

Figure 4.25 Create an access hole by drilling a circular pattern of small holes, then punch out the door material within the circle.

Figure 4.26 One of the fastest ways to force a door is to saw it in half from top to bottom.

Battering Swinging Doors

One of the oldest methods of forcing a swinging door is to break it in. Firefighters have often used body force or striking and battering tools such as sledges and battering rams to break a locked door out of its frame. This text does not advocate any method of forcing a door using the body as a striking instrument. The human body is not designed for such use, and injury is likely to occur when a well-secured door will not yield to bodily impact. There are many tools superior to the human body for forcible entry.

Choose a tool that is heavy enough to transmit the most force with the least effort. A heavy sledge or battering ram is clearly superior to smaller versions of the same, particularly when the door is well constructed. Heavy tools can generate tremendous force when used in the proper manner and should always be chosen for this type of application.

Forcing a door by striking or battering almost always means partial or total destruction of the door or frame. Cheaper, lightweight doors are best for this type of entry because the chances of success are much greater than with well-constructed doors. In addition, the cost of replacement is much less for a cheaper door, such as the hollow-core type.

We usually think in terms of in-swinging doors when preparing to batter a door; although in some cases, out-swinging doors can also be battered. The impact point for the striking tool should be either near the latch (Figure 4.27), or on the locking device itself. This maximizes the force where it is most likely to spring the door free. If a door is struck at its center, the force is often transmitted outward through the door with little effect on the locking

Figure 4.27 One of the most effective places to strike a door with a battering ram is near the latch.

device or hinges. On cheaper doors, battering the door center may result in creating a hole, but this at least allows access to the interior lock release.

Out-swinging doors can best be breached if they are battered at the locking mechanism. Destroying the lock may be the only way to release the latch and pull the door open. It is possible to break the stop loose by battering a stopped door, but usually only if the door frame is of inferior wood construction.

FORCING SLIDING DOORS

Sliding doors travel in the same plane to the right or left of their opening. They usually slide on small rollers or guide wheels on a metal track. Ordinary sliding doors travel into a partition or wall when opened, a common arrangement for interior openings. These doors can be forced in the same way as hinged doors, except that they must be pried straight back from their locks.

A sliding patio door is one of the most difficult doors to force without causing significant damage. Consisting of heavy-duty, full-panel glass set in a metal or wood frame, the patio door usually slides over a stationary glass panel. Some older patio door units have the stationary panel set in the frame with screws exposed to the exterior. Removing the screws allows removal of the panel, thus permitting entrance through that side of the frame.

Regardless of the door design, avoid breaking the glass panels to gain entry. Older doors may be composed of plate glass, which breaks into jagged, knife-like shards when broken. Newer doors contain tempered glass in single or double sheets. The double-sheeted door, called "thermopane" (two panels of glass with a hermetically sealed air space between), is a relatively expensive door to repair if the glass is broken.

Patio doors can be forced open by inserting a prying tool between the frame and the door near the lock and prying the door away from the frame. The objective here is to break the latch out of the strike. Remember that prying improperly can do more damage than breaking the latch.

Another method involves lifting the sliding door so that the latch disengages from its strike. This is accomplished by inserting prying tools under the door, prying upward until the door touches the top of the frame, then sliding the door open. This method is only possible if the latch engages the strike bar in a downward direction, and if there is enough space inside the frame to allow the latch to clear the strike during the lifting process.

Patio sliding doors are sometimes barred or blocked by a metal rod or other device. Such a device, which usually can be seen from the outside, practically eliminates any possibility of forcing. The only way to disengage this type of device is by inserting a thin-bladed tool, such as a long shove knife, between the door and the stationary panel and manipulating the bar out of its track.

FORCING FOLDING DOORS

Folding doors are not commonly used as exterior doors, but occasionally are encountered in commercial occupancies where space restrictions require a door that does not block the exit passage when opened. Prying is one of the best methods to force a folding door if through-the-lock entry is not possible. Commercial doors are often mounted in a double bi-folding arrangement and may be operated electrically. When the building is locked, the doors are especially resistant to forcing because the activating source is de-energized. This, in effect, locks the door into a closed position because the articulating hardware is fully extended and locked into place. In this case, it is advisable to seek another point of entry.

FORCING OVERHEAD DOORS

Doors that open by moving overhead can be classified into overhead rolling, folding, and slab doors. Overhead rolling doors are constructed of narrow interlocking slats that roll up into a barrel when the door is opened. Overhead folding doors are composed of metal, fiber glass, or wood-framed sections that contain wood or glass panels. Overhead slab doors are composed of a single rigid unit that swings open with the help of large springs. Slab doors may contain small glass or masonite panels that can be easily replaced if broken during forcible entry.

Manually operated overhead doors often have a locking handle, usually in the center, that controls bars on both sides of the door (Figure 4.28). In the locked position, the bars seat into strikes in the frame. Such a door can be forced by prying up at the bottom of the door with a prying tool, but less damage occurs and less time is used by knocking out a panel or cutting an access hole near the center (Figure 4.29) and turning the handle from the inside.

Some overhead doors are locked with a padlock or steel pin through a hole at both ends of a bar that may be attached to the door frame or track. By knocking out a panel as described above, a firefighter can climb through the opening, cut and remove the locking device with bolt cutters, and open the door.

Electrically operated doors (often remote-controlled) are usually locked by a mechanism in the drive motor when they are in the closed position. These doors have an emergency release so that they can be opened manually from the inside in the event of a power failure (Figure 4.30). If a narrow tool can be inserted at the top of a closed door, it may be possible to trip the release, allowing the door to be opened manually. If this is not possible, cut or break out a panel, if the door is so constructed, to gain access to the release.

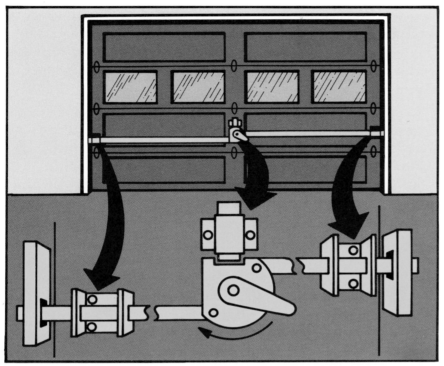

Figure 4.28 Some manually operated overhead doors have a center-mounted door handle that controls locking bars, which slide into the frame or track.

Figure 4.29 One of the fastest ways to unlock an overhead door with a center-mounted handle is to break out a panel and reach inside to operate the handle.

Figure 4.30 Some electrically operated doors have a manual release for the locking mechanism.

Overhead rolling doors can be made of fabricated steel, aluminum, fiber glass, or wood. They are found in garages and warehouses, in service entrances to mercantile and industrial buildings, and as security covers on mercantile buildings. All are similar structurally, and they can be operated mechanically, manually, or electrically. The barrel on which the door is wound is usually turned by a set of gears near the top of the door inside the building. This makes the door exceptionally difficult to force. During pre-incident planning, firefighters should try to learn the best way to operate the door. If firefighters cannot operate the door during a fire, they can make a large V-cut with a saw (Figure 4.31). Because of the cost of repair or replacement of a rolling door, consider entering at another point and operating the door in a normal manner from the inside.

An overhead rolling grill can be forced in a manner similar to that used for overhead doors, usually with a power saw. As with overhead doors, it is sometimes better to make entrance at another point and operate the grill in a normal manner from inside. Be sure to inspect these grills during pre-incident planning to note the proper way to open them.

Overhead slab doors are made of metal, fiber glass, or wood, and are similar to overhead folding doors. Unless slab doors have glass windows, it is practically impossible to reach the latch on the inside. Instead, use a bar to try to pry them out at each side near the bottom, bending the lock bar enough to pass the strike. CAUTION: All overhead doors should be blocked open (up) to prevent injury to firefighters if the control device should fail. One method is to use a "C" clamp or vise grip pliers to block the track (Figure 4.32). Another method is to block an open door with a pike pole or a ladder.

Figure 4.31 Make a large V-cut with a saw to open an overhead door that has interlocking slats. *Courtesy of Los Angeles City Fire Department.*

Figure 4.32 To prevent an overhead door from accidentally closing, clamp a pair of vise grip pliers to the track.

FORCING REVOLVING DOORS

A revolving door has quadrants, or wings, that revolve around a center shaft. Each wing is held in place by hardware at the top and bottom. The door assembly revolves in a metal or glass housing that is open on both sides and through which pedestrians travel as the door turns. In order to prevent jamming in case of panic, the door mechanism is usually collapsible and thus panic proof. When any two wings are pushed in opposite directions, the wings book fold to a stationary-open position. All revolving doors do not collapse the same way, so it is good policy to collect such information when making pre-fire surveys.

The revolving door, when locked, presents a formidable forcible entry problem because the locking mechanism is usually only accessible from the inside. This type of door is intended to be locked by someone who leaves by another door after securing the building. Revolving doors are expensive, thus it is best to avoid damaging them (Figure 4.33). The best tactic to employ when deciding how to force a revolving door is...... **DON'T!** Choose another entry point.

Unlocked revolving doors present an obstacle to convenient entry by firefighters with equipment. It is difficult to lay hose through a normally functioning revolving door.

There are three types of collapsing revolving doors:

- Panic-Proof
  The panic-proof collapsible mechanism is designed to book fold when two wings are pushed in opposite directions (Figure 4.34).

Figure 4.33 Repair costs on damaged revolving doors are considerable. DO NOT FORCE A REVOLVING DOOR UNLESS IT IS ABSOLUTELY NECESSARY.

Figure 4.34 To book fold the wings of an unlocked revolving door, push two wings in opposite directions.

- Drop-Arm

  The drop-arm mechanism has a solid arm passing through a pawl in one of the doors. To collapse the mechanism, press the pawl to disengage it from the arm and push the wing to one side.

- Metal-Braced

  The metal-braced mechanism is held in place by wing arms that resemble gate hooks with eyes. To collapse the mechanism, lift a wing arm and fasten it back against the fixed wing. The arms are on both sides of these doors. In most cases, the pivots are made of cast iron and can be broken by forcing the door with a bar at the pivots.

FORCING FIRE DOORS

Fire doors usually do not need to be forced because they are designed to protect from fire rather than to prevent entry. Fire doors that close automatically need only be pushed open to gain entrance. In some cases, such doors become wedged closed by excessive heat that swells the door in its frame. If this happens, it is necessary to pry the door open with suitable forcible entry tools. It is important to remember that whenever a fire door is closed, fire could be on the other side. Have hoselines charged and in place when forcing a fire door.

Fire doors on exterior openings may be found in such places as smoke towers and stairwells. Stairwell fire doors are often locked to prevent entry from the outside, but allow occupants unrestricted egress from the corridors. Interior stairwell doors lock behind occupants who move from the building corridors into the stairway, thus prompting egress toward an exterior exit door at ground level. This arrangement prevents occupants from re-entering corridors that may be filled with fire or smoke.

Firefighters should remember, however, that unless such doors are unlocked or blocked open, they could also be denied access to areas for rescue and fire fighting. While it is inadvisable to wedge these doors open, the locking mechanism can be disarmed by key or by placing a device such as a rubber band over the latch (Figure 4.35). This allows the door to close, preventing contamination of the stairwell by smoke and fire, but permits two-way travel through the door.

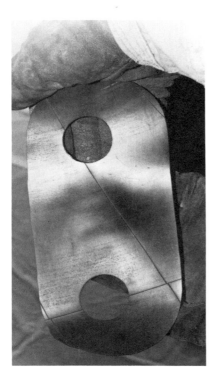

Figure 4.35 To keep a swinging fire door from locking behind you, place a rubber band over the latch.

Answer each of the following questions in a few words or short phrases:

1. Once a decision has been made to force a door to enter a locked building, what criteria should be used to determine which door should be forced?

   A. _____

   _____

   B. _____

   _____

   C. _____

   _____

   D. _____

   _____

2. Of the four criteria listed above, which is the most important? Why?

   _____

   _____

   _____

   _____

3. It has been decided to enter a building through the rear swinging door. What method of entry should be attempted *first*?

   _____

   _____

   _____

   _____

4. When sizing up a door before forcing it, what questions about the door should be answered?

   A. _____

   _____

   B. _____

   _____

   C. _____

   _____

   D. _____

   _____

   E. _____

   _____

5. Two doors are identical in design and construction. The only difference is that one opens outward and the other opens inward. Which is easier to force open? Why?

_____

_____

_____

6. When attempting to batter open a swinging door with a heavy sledge, where should the impact point be?

_____

_____

_____

7. What are three types of overhead doors?

A. _____

B. _____

C. _____

8. You are attempting to force open an overhead door with glass panels. What is the best way to gain entry through this door?

_____

_____

_____

Determine whether the following statements are true or false. If false, state why:

9. Forcible entry should be accomplished with minimal damage, which means that the operation must be done both slowly and carefully.

☐ T   ☐ F   _____

_____

10. A wood door may require a hard shove to open because, under certain conditions, it tends to swell in its frame.

☐ T   ☐ F   _____

_____

11. Before opening a door to a building that may contain fire, feel the door with a bare hand for heat.

☐ T   ☐ F   _____

_____

12. Swinging doors should be pried open only on the lock side, not on the hinged side.

☐ T   ☐ F   _____

_____

13. In-swinging doors cannot be pried open because the stop or rabbet blocks access to the door edge and latch.

☐ T   ☐ F   _____

_____

14. Unlike swinging doors, sliding doors cannot be pried open.

☐ T   ☐ F   _____

_____

15. When a patio door is blocked by a wooden or metal rod, the best method of gaining entry is by breaking the glass.

☐ T   ☐ F   _____

_____

16. After an overhead door has been forced open, it should be blocked open in case the control device fails.

☐ T   ☐ F   _____

_____

17. A revolving door should be forced open only when there is no other way to gain entry to the building.

☐ T   ☐ F   _____

_____

18. Fire doors on stairwells that allow exiting to the outside but are locked to prevent entrance from the exterior should be propped open during fire fighting operations to permit entrance by firefighters.

☐ T   ☐ F   _____

_____

Select the choice that best completes the sentence or answers the question:

19. Firefighters should check for signs of a backdraft
  A. before forcing open a door
  B. during forcible entry
  C. after opening a door
  D. while searching the building

20. The first thing to check on a swinging door is
    A. the type of lock in the door
    B. the direction in which it opens
    C. whether it is solid or hollow core
    D. in what type of frame it is mounted

21. When prying open a swinging door on the lock side, hand prying tools should be inserted between the door and frame
    A. near the latch or handle
    B. at the midpoint of both the bottom and top panels
    C. first near the bottom, then near the top of the door
    D. any of the above, depending on door construction

22. The objective in prying open double swinging doors is to
    A. allow access to the inside bars
    B. separate one door from the other
    C. separate the anchor door from the door frame
    D. slip latches from their strikes

23. Which tool can be used to cut a hole in a door to gain access to the lock?
    A. Hole saw
    B. Drill
    C. Power saw
    D. Any of the above

Match the tool in the left column with the recommended usage in the right column:

	Tool	Usage
____	**24.** Pry-axe	A. Spreading a door frame
____	**25.** Rabbit tool	B. Prying an out-swinging wood door
____	**26.** Detroit door opener	C. Cutting open an overhead rolling door
		D. Prying an in-swinging door in a metal frame

Lock
Assemblies

5

NFPA STANDARD 1001
FORCIBLE ENTRY
Fire Fighter II

4-2 Forcible Entry

4-2.1   The fire fighter shall identify materials and construction features of doors, windows, roofs, floors, and vertical barriers and shall define the dangers associated with each in an emergency situation.

Chapter 5
Locking Assemblies

Through-the-lock forcible entry is one of the best ways to gain access to the inside of a locked building with the least amount of damage. This is especially important when firefighters must get inside a building for purposes other than fire fighting, such as medical aid calls, smoke or odor investigations, and "alarm sounding" calls. In each of these cases, it is difficult to justify damage if the call turns out to be a false alarm or a needless call. Through-the-lock entry allows firefighters to disable locking devices with little or no damage.

To successfully disable all types of locks encountered, preparation is required in three areas:

- Locking device type and function
- Through-the-lock entry methods
- Pre-incident planning

Pre-incident planning is listed last because the most effective planning is accomplished by personnel who are well trained in the first two areas. This training enables them to identify the types of locks found in each occupancy and to plan the best forcing method, based upon a knowledge of the construction and function of each lock.

As with any means of forcible entry, a thorough knowledge of the components involved is essential to a speedy and efficient operation (Figure 5.1 on next page). Locking devices come in a wide range of designs and styles that can be confusing when sizing up the situation. Locks that are apparently similar may operate very differently. Each, therefore, must be approached in a different manner.

Recognizing the basic categories of locks is critical to an understanding of lock function. This knowledge aids in the size-up

Figure 5.1 A thorough knowledge of locks is essential to performing speedy, efficient through-the-lock entry.

process, because forcing methods are based upon the operation of the internal mechanism of each type of lock.

The American National Standards Institute (ANSI) and most lock manufacturers divide locks into five categories:

- Bored (Cylindrical)
- Mortise
- Rim
- Preassembled (Unit)
- Exit Device

LOCK TERMINOLOGY

To better understand how locks are constructed and how they operate, knowledge of the terminology used to describe lock components is essential. A few basic terms and their definitions are provided for this purpose. A more complete list is found in the Glossary section of this manual.

ASTRAGAL — A molding that covers the narrow opening between adjacent double doors in the closed position.

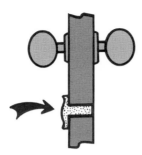

AUXILIARY DEADBOLT — A deadbolt bored lock; also called "tubular deadbolt."

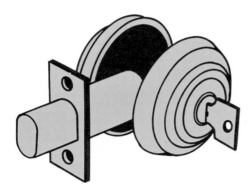

AUXILIARY LOCK — A lock added to a door to increase security.

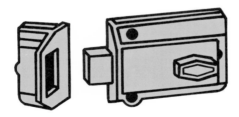

BACK PLATE — The plate used with a rim lock to secure the lock cylinder to the door.

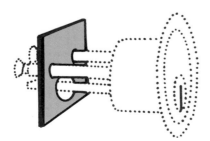

BORED LOCK — A lock installed within right-angle holes bored in a door; also called "cylindrical lock."

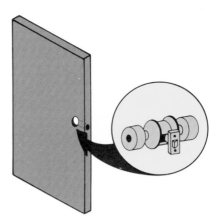

BRACE LOCK — A rim lock equipped with a metal rod that serves as a brace against the door.

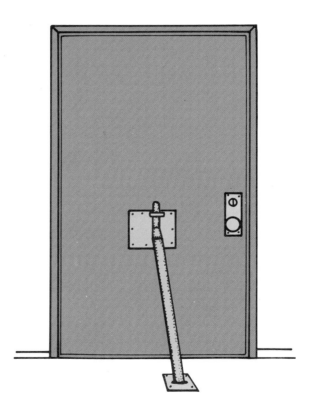

CAM — The part of a mortise lock cylinder that moves the bolt or latch as the key is turned.

CASE (LOCK) — The housing for any locking mechanism.

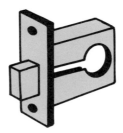

CYLINDER (LOCK) — The component of a locking mechanism that contains coded information for operating that lock, usually with a key.

CYLINDER GUARD — A metal plate that covers a lock cylinder to prevent forceful removal.

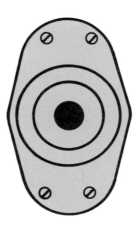

CYLINDER SHELL — The external case of a lock cylinder.

DEADBOLT — The movable part of a deadbolt lock that extends from the lock mechanism into the door frame to secure the door in a locked position.

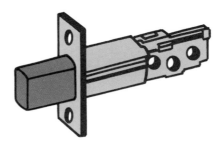

DEAD LATCH — A sliding pin or plunger that operates as part of a dead locking latch bolt; also called "anti-shim device."

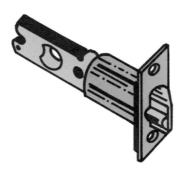

DROP BAR — A metal or wooden bar that serves as a locking device when placed or "dropped" into brackets across an in-swinging door.

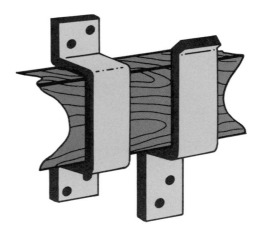

EXIT DEVICE — A locking assembly designed for panic exiting that unlocks from the inside when a release mechanism is pushed; also called "panic hardware."

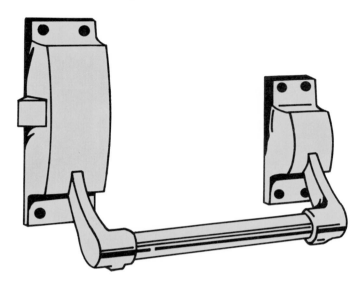

FLUSH BOLT — A locking bolt that is installed flush within a door.

HASP — A fastening device consisting of a loop eye or staple and a slotted hinge or bar; commonly used with a padlock.

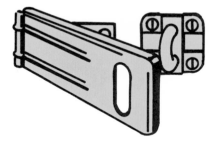

JIMMY — To pry apart, usually to separate the door from its frame to allow the latch or bolt to clear its strike.

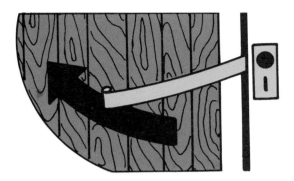

JIMMY-RESISTANT LOCK — An auxiliary lock having a bolt that interlocks with its strike and thus resists prying; also called "vertical deadbolt" or "interlocking deadbolt."

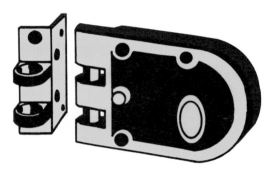

KEY — A device that, when inserted into a key plug, causes the internal pins or disks to align in a manner that allows the plug to turn within the cylinder; a device that allows the operator to lock and unlock a locking mechanism.

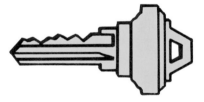

KEY-IN-KNOB LOCK — A lock in which the lock cylinder is within the knob.

KEY PLUG — The part of a lock cylinder that receives the key; also called the "cylinder plug."

KEYWAY — The opening in a cylinder plug that receives the key.

LATCH — The spring/loaded part of a locking mechanism that extends into a strike within the door frame.

LATCH BOLT (DEAD-LOCKING) — A latch with a shim or plunger that causes the latch to operate in a manner similar to a deadbolt; the latch plunger prevents "loiding" of the latch.

LOCK — A device for fastening, joining, or engaging two or more objects, such as a door and frame, together.

LOCK MECHANISM — The moving parts of a lock, which include the latch or bolt, lock cylinder, and articulating components.

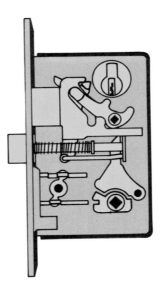

3-89

LOIDING — The method of slipping or shimming a spring latch from its strike with a piece of celluloid (credit card).

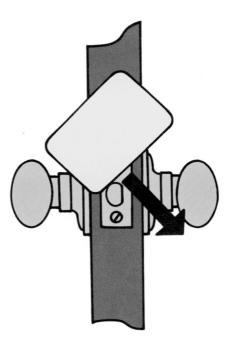

MORTISE CYLINDER — A lock cylinder for a mortise lock.

MORTISE LOCK — A lock mortised into a door; also called "box lock."

MULLION — A center post, sometimes removable, in a double door opening.

MULTIBOLT LOCK — A high-security lock that uses metal rods to secure the door on all sides.

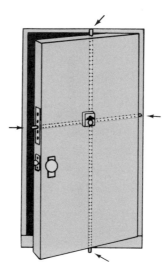

NIGHT LATCH — A button on a rim lock that prevents retracting the latch from the outside.

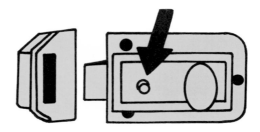

PADLOCK — A detachable, portable lock with a hinged or sliding shackle.

PIVOTING DEADBOLT — A lock having a deadbolt that pivots 90 degrees, designed to fit a narrow-stiled door.

PREASSEMBLED LOCK — A lock designed to be installed as a complete unit (requiring no assembly) within a door; also called "unit lock."

RIM CYLINDER — A lock cylinder for a rim lock.

RIM LOCK — A type of auxiliary lock mounted on the surface of a door.

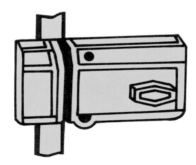

SHACKLE — The hinged part of a padlock.

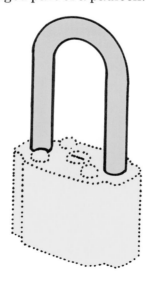

SHEAR LINE — The space between the shell and the plug of a lock cylinder obstructed by tumblers in the locked position.

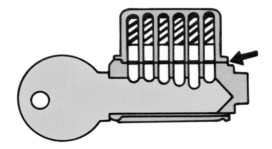

SHOVE KNIFE — A tool for "loiding" a latch.

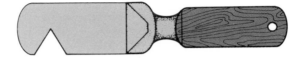

SKELETON KEY — A key for a warded lock.

STEM — The part of a lock cylinder that activates the bolt or latch as the key is turned; also called "tailpiece."

STRIKE — The metal plate mounted in the door frame that receives the latch or deadbolt.

SURFACE BOLT — A sliding bolt installed on the surface of a
    door.

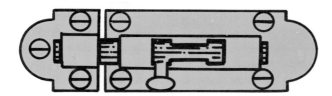

THUMBTURN — A part of the lock, other than the key or knob,
    used to lock and unlock the door.

TUMBLER — A pin in the tumbler-type of lock cylinder.

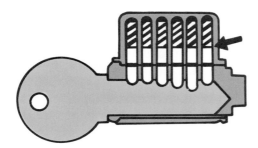

WARDED LOCK — A simple type of mortise lock that requires a
    skeleton key to open.

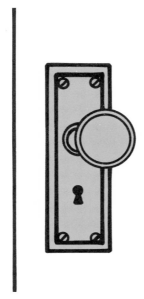

BORED (CYLINDRICAL) LOCK

Bored locks are so named because their installation involves boring two holes at right angles to one another: one through the face of the door to accommodate the main locking mechanism, and the other in the edge of the door to receive the latch or bolt mechanism (Figure 5.2). When these two mechanisms are joined into an assembly within the door, they form the complete latching or locking mechanism. These locks may serve as the primary locking device or as an auxiliary lock.

Key-in-Knob Lock

Some bored locks have lock cylinders in the knobs (Figure 5.3). Commonly called key-in-knob locks, they are made in three weights: light, standard, and heavy.

NOTE: Although the term "key-in-knob" usually refers to bored locks, this is not always the case. Some locks that contain lock cylinders in the knob are mortise locks, while others are pre-assembled locks.

The key-in-knob lock has a keyway in the outside knob; the inside knob may contain either a keyway or a button. The button may be a push button or a push-and-turn (thumbturn) button. Some designs contain no locking mechanism in the inside knob. Key-in-knob locks are equipped with a latch mechanism that is locked and unlocked by both the key and, if present, by the knob button. In the unlocked position, a turn of either knob retracts the spring-loaded beveled latch bolt, which is usually no longer than 3/4-inch (19 mm).

Figure 5.2 A bored (cylindrical) lock requires two holes in the door — one bored into the door face, the other bored at a right angle through the door edge.

Figure 5.3 Some bored locks have lock cylinders in the knob and they are called "key-in-knob" locks.

Well-constructed latching locks have a spring-loaded plunger fitted against the latch bolt (Figure 5.4) to prevent its forced retraction by a tool. When pressure is applied to the beveled face of the latch bolt to push it out of the strike, the plunger (sometimes called a dead-latch or anti-shim device) causes the latch bolt to "lock" into place. The entire mechanism is known as a "dead locking latch bolt."

Because of the relatively short length of the latch, key-in-the-knob locks are one of the most vulnerable to prying operations. If the door and frame are pried far enough apart, the latch clears the strike and allows the door to swing open.

Figure 5.4 Well-constructed latching locks have a spring-loaded plunger (anti-shim device) fitted against the latch; its purpose is to prevent "loiding" of the latch.

HOW THE KEY-IN-KNOB LOCK WORKS

The bored type of key-in-knob lock is composed of two knobs that are usually connected by two screws (Figure 5.5 on next page). A stem (also called a tailpiece) is attached to, and operated by, the knobs. This stem runs through the latch mechanism, mounted at right angles to the knob assembly. The stem is the component that retracts the latch. In the unlocked position, either knob retracts the latch when turned in either direction. Turning a knob turns the stem, which in turn retracts the spring-loaded latch inside the tube of the latching mechanism (Figure 5.6 on next page).

The locking mechanism is activated when the key is turned or the button in the inside knob is either pushed or turned, depending upon its design. This action prevents the knob from turning, usually by pushing the stem into a position that prevents it from turning within the latch assembly.

Auxiliary Deadbolt

Auxiliary deadbolts, often called tubular deadbolts, are bored deadbolt locks that are usually found installed above another lock. These simply constructed auxiliary locks provide

Figure 5.5 The bored key-in-knob lock is composed of two knobs connected with two screws. A stem (tailpiece) between the knobs activates the latching mechanism, which is mounted in the door edge.

Figure 5.6 Turning the knob of a key-in-knob bored lock rotates its tailpiece, which in turn retracts the latch from its strike.

excellent security. Consisting of a cylinder installed on one side of a door, with a cylinder or thumbturn plate on the opposite side (Figure 5.7), a deadbolt projects laterally into the strike and door frame. The locking mechanism operates with a key in the outside cylinder and with either a key or a thumbturn on the inside.

As with key-in-knob locks, auxiliary deadbolts may be of light, medium, or heavy-duty construction. The heavy-duty locks are very difficult to force, primarily because they contain large case-hardened screws and other heavy-duty components that resist prying and cutting operations. Deadbolts are 1 to 2 inches (25 to 50 mm) long and slide within a hollow tube to lock the door. Some models feature a floating, case-hardened steel rod inside the deadbolt body. Attempting to cut the deadbolt with a saw causes the rod to spin inside the bolt when the saw blade touches it, halting progress through the remainder of the deadbolt body (Figure 5.8).

Figure 5.7 An auxiliary (tubular) deadbolt (upper lock) consists of a cylinder on the outside of the door and either a key cylinder or a thumbturn on the inside of the door. A deadbolt of up to 2 inches (50 mm) in length is mounted in the door edge.

Figure 5.8 Some auxiliary deadbolts have a case-hardened steel rod that, if the deadbolt body is cut, spins when it comes into contact with a saw blade. This effectively prevents further penetration of the deadbolt body.

HOW THE AUXILIARY DEADBOLT WORKS

The auxiliary deadbolt is operated by turning a key in the outside cylinder, which contains the locking mechanism, or by an interior thumbturn or key. Turning the key or thumbturn rotates a tailpiece that projects from the back of the key cylinder (Figure 5.9). The tailpiece rotation within the lock mechanism, in turn, causes the deadbolt to slide in its tubular housing.

Figure 5.9 Turning either the key or the thumbturn of an auxiliary deadbolt lock retracts the deadbolt from its strike by rotating a short stem that slides the deadbolt within its tubular housing.

MORTISE LOCK

Mortise locks, sometimes called box locks, are among the oldest locks still in use today. The basic design is over 150 years old, and has been changed only to permit the use of tumbler-type key cylinders and deadbolts to make the lock more secure. Earlier locks, still found on some older residences, have a "peeping tom" keyhole for a skeleton key (Figure 5.10). This is known as a "warded lock" and few are still in use today because they can be easily picked. The earlier locks also employed only a spring-loaded latch of short length (no more than 1/2-inch [13 mm]), which allowed the door to be easily jimmied.

Figure 5.10 One of the oldest styles of mortise lock requires only a skeleton key to activate its locking mechanism.

Newer mortise locks (Figure 5.11) not only have latches, but also deadbolts up to 1 inch (25 mm) long. Key cylinders are installed in the lock by a threaded port in the lock housing and are further anchored by a set screw that bores into the side of the cylinder case. The cylinder case is usually made of brass or another soft metal. Because the cylinder is installed in this way, the mortise lock is vulnerable to through-the-lock entry.

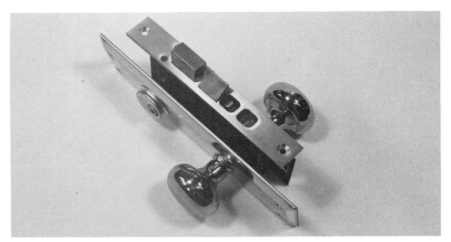

Figure 5.11 Modern mortise locks have not only latches, but also deadbolts up to 1-inch (25 mm) long.

How the Mortise Lock Works

A typical mortise lock is pictured in Figure 5.12. The lock illustrated is the type that can be locked in two different ways: (1) with the latch only, or (2) with the deadbolt and latch.

A — UNLOCKED The lock in the upper left section of the illustration is shown in the unlocked position. In this position, a turn of the doorknob retracts the spring-loaded latch from the strike, allowing the door to swing free. The deadbolt is nonfunctional in the retracted position.

B — LATCH LOCKED The latch locking button in the faceplate has been pushed in to engage the door handle pivot. This locks the latch so that the doorknob cannot turn to retract the latch.

C — LATCH DISENGAGED To unlock the door from the outside, the key is turned a quarter-turn clockwise, which rotates the cam at the back of the cylinder down against a rocker arm. The arm pivots to push a rod against the latch back plate, which in turn retracts the latch.

D — DEADBOLT AND LATCH LOCKED When the key is rotated three-quarters of a turn counterclockwise, the cam engages the deadbolt throw, extending the deadbolt into the strike. The deadbolt back plate butts against the back of the latch at the same time, locking it into place.

The key must be rotated another quarter-turn counterclockwise to remove it from the cylinder.

E — DEADBOLT AND LATCH UNLOCKED When the key is rotated three-quarters of a turn clockwise, the cam engages the deadbolt throw, retracting the deadbolt into the lock housing. The latch is released at the same time. The door can be opened normally as long as the latch lock is also retracted.

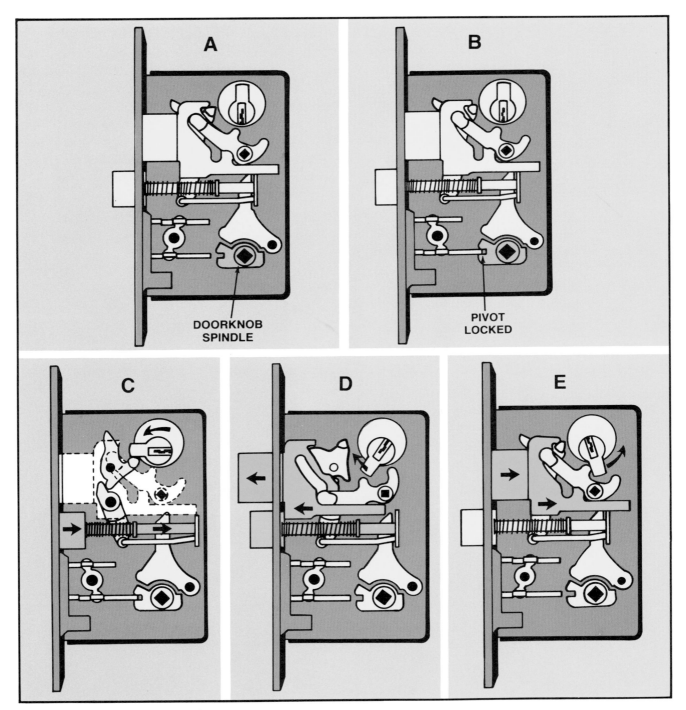

Figure 5.12 How the mortise lock works.

Figure 5.13 A rim lock is usually installed above an existing lock to add security.

RIM LOCK

The rim lock is one of the most common locks in use today. It is best described as being surface-mounted and for this reason is used as an add-on lock for doors that have other types of locks. This lock is found in all types of occupancies, including houses, apartments, and some commercial buildings.

The rim lock is usually installed above an existing lock as an auxiliary security device (Figure 5.13). It can be identified from the outside by a cylinder that is recessed into the door in a bored hole. In addition to the external cylinder, the rim lock consists of a latching mechanism fastened to the inside of the door and a strike mounted on the edge of the door frame. The cylinder and latching mechanism are connected by two screws and the cylinder is held in place within the bored hole by a "rim" or ring of metal (Figure 5.14). When the securing screws projecting from the locking mechanism are set, the cylinder tightens against the rim, anchoring the entire assembly. A tailpiece (stem) projects from the back of the lock cylinder to activate the locking mechanism.

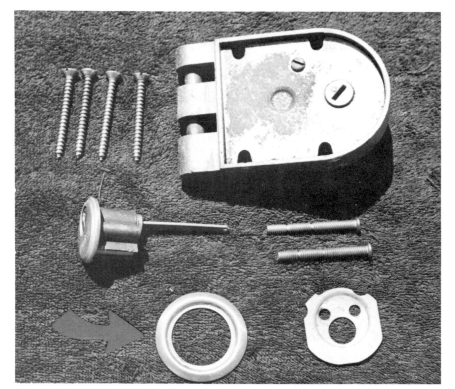

Figure 5.14 Two screws connect a rim lock cylinder to the latching mechanism, which is mounted on the inside face of the door. The cylinder is held in place by a metal ring, or "rim."

Rim locks are available in three configurations:

- Latch
- Deadbolt
- Interlocking deadbolt

The latching type of rim lock (Figure 5.15) has a 1/2-inch (13 mm) spring-loaded, beveled latch that can be locked into the open or closed position with a "night latch" button. The deadbolt model (Figure 5.16) has a deadbolt, 1/2-inch to 1-inch (13 mm to 25 mm) long that can be locked from the inside by rotating a thumbturn. This lock is far more resistant to prying operations than the latching type because the deadbolt is longer.

A more secure version of the deadbolt lock is the interlocking deadbolt (Figure 5.17), also known as the "vertical deadbolt" or as the "jimmy-proof lock." The lock is extremely effective because the locking mechanism (door-mounted) interlocks the strike (frame-mounted), and virtually eliminates the possibility of "jimmying" the door.

Figure 5.15 A latching rim lock usually has a "night latch" button for extra security.

Figure 5.16 The deadbolting rim lock is more pry-resistant than the latching rim lock.

Figure 5.17 The interlocking deadbolt is the most pry-resistant rim lock (lock is shown from the back side). See also Figure 5.13.

How the Rim Lock Works

A rim lock operates from the outside with a turn of a key in the lock cylinder, which is mounted in the outside door face. Turning the key causes a tailpiece to rotate within the locking mechanism to extend and retract the latch or deadbolt. A thumbturn on the interior face of the lock accomplishes the same purpose. A "night latch" type of rim lock does not unlock with a key from the outside if the lock has been "night latched" from the inside.

The interior mechanism of the interlocking deadbolt is the same as that of other types of rim locks. Turning a key or thumbturn rotates the tailpiece, which causes a two-pronged bolt to slide into vertically aligned recesses within the strike (Figure 5.18); this feature interlocks the locking mechanism on the door with the frame-mounted strike.

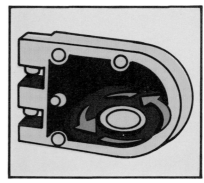

Figure 5.18 Turning the key or thumbturn of an interlocking deadbolt causes the two-pronged bolt to slide into vertically aligned recesses within the strike.

PREASSEMBLED (UNIT) LOCK

The preassembled lock (Figure 5.19) is a heavy-duty lock designed for easy installation, requiring only a U-shaped cutout in the door edge (Figure 5.20). The entire locking mechanism is preassembled into a single "unit" so that the installer does not have to align separate lock components within holes or recesses in the door frame, as with bored or auxiliary locks. The design virtually eliminates the problem of misalignment that often occurs with lock installations.

This type of lock, commonly found in commercial occupancies, is highly resistant to forcing operations because it is composed of high-quality metal components. In most designs, the locking mechanism is located in the knob and thus resembles the common "key-in-knob" lock. It can be recognized as a preassembled unit, however, by the presence of what appears to be a metal faceplate behind the knob. The faceplate is actually one side of the housing that encases the lock mechanism.

Figure 5.19 A preassembled (unit) lock is a heavy-duty lock, usually found in commercial buildings.

Figure 5.20 The preassembled lock requires only a U-shaped cutout within the door edge. Its design was intended to make installation easier.

How the Preassembled (Unit) Lock Works

Operating in a manner similar to other locks, some preassembled locks use a stem to retract the spring-loaded latch when the knob is turned in either direction (as do standard key-in-knob locks). Other preassembled locks use a cam device (similar to mortise locks), which rotates to retract the latch as the knob turns.

The assembly locks when a key is turned in the exterior knob keyway or when a button in the interior knob is either pushed or turned. This action "locks" the knob by moving the internal stem or cam into a position that prevents it from turning within the latch assembly.

EXIT DEVICE

In buildings such as hospitals, schools, and theaters, or any occupancy where a life safety hazard exists, a special type of lock is used to secure exterior doors. This type of lock is simply called an "exit device."

Exit devices are operated by pushing a bar, rather than by turning a knob. This causes the lock to release whenever force is applied to its "panic hardware" in the direction of exit, as happens when people are pushed against the door in a panic situation.

Exit devices are available in four configurations:

- Rim
- Mortise
- Surface Vertical Rod
- Concealed Vertical Rod

Rim and mortise devices are installed on single swinging doors; vertical rod devices are used on double swinging doors.

Rim exit devices are add-on locks. They are often found in occupancies that business owners have retrofitted to meet fire code requirements for exiting. The entire assembly is surface mounted on the face of the door (Figure 5.21).

Figure 5.21 A rim exit device is mounted on the surface of a door, often as a retrofit lock.

The mortise exit device is installed as an integral part of the door, with the locking mechanism mortised into the door (Figure 5.22). This device is usually found in newer buildings in which doors have been custom fitted during initial construction.

The vertical rod exit device is the only type of exit lock that locks both doors at the top and bottom. Surface-mounted vertical rod devices are usually added to a door some time after the door is installed (Figure 5.23). The *concealed* vertical rod device is found wherever other mortised locks are used, usually in newer installations. In this case, the rod mechanism is encased within the door and is not visible.

When double swinging doors are equipped with exit devices, there is usually some type of hardware mounted to cover the center joint between doors. This cover may be an attached strip of molding called an astragal that is fastened to one leaf. Some installations have a vertical bar that is installed in the center of the opening. This bar, known as a mullion, is used when doors have rim-type exit devices that must latch into a center post (Figure 5.24). In some openings, the mullion is removable so that it will not hinder the passage of people in a panic situation.

Most exit devices have a lockdown, or "dogging" feature that causes the latch bolts to remain in a retracted position so that the door can be operated in a push-pull manner. This feature is used to "unlock" the door so that traffic can pass in both directions through the door without having to activate the exit device.

Newer designs of exit devices have low profile crossbars that operate horizontally, rather than vertically downward and outward (Figure 5.25). This change in design was made not only for aesthetic purposes, but also to make it more difficult to manipulate the locked hardware from outside the door. With older designs it is possible to slip a wire in from the outside to engage and operate the crossbar. The new design of crossbar virtually eliminates this forcing method.

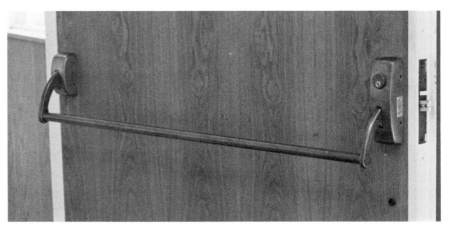

Figure 5.22 A mortise exit device is installed as an integral part of the door.

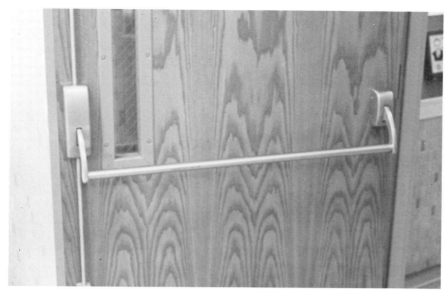

Figure 5.23 A surface-mounted, vertical rod exit device locks a door both at the top and bottom.

Figure 5.24 Some double doors have a center post, or mullion, in the opening.

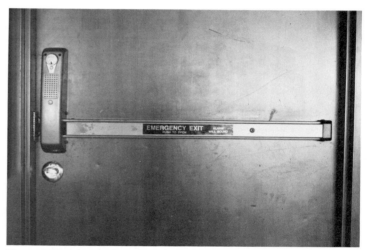

Figure 5.25 A low-profile exit device.

How the Exit Device Works

All exit devices have some type of actuating bar mounted horizontally across the door surface at about waist height. A few special installations allow a paddle rather than a crossbar (Figure 5.26 on next page), and this is found on the side of the door closest to the latch. Both devices actuate the locking mechanism when pushed, which in turn retracts either a latch or vertical rods so that the door swings freely.

Figure 5.26 Some exit devices have paddle-like push arms, that may activate an alarm when they are pushed.

KEY CYLINDER

The heart of any lock or locking mechanism is the key cylinder. Conventional cylinders are made of two basic parts: a shell and a key plug (Figure 5.27). The shell, or exterior part of the cylinder, is made to hold the key plug. The key plug is the part of the lock that accepts the key. After the two components are assembled, a precisely aligned row of holes is drilled through the side of the shell into the key plug to accept pins (tumblers). A set of two or three pins and a spring is loaded into each of the holes and the holes are sealed. The springs put downward tension on the pins so that they bridge the bottom edge of the holes in the shell and the top edge of the holes in the plug. In this position, the pins prevent the plug from rotating within the cylinder and thus "lock" the cylinder. The lengths of the pins in the key plug correspond to indentations in the key. When the key is inserted into the plug, the pins are pushed up against the springs in correct alignment to form a shear line, allowing the key to turn and rotate the plug

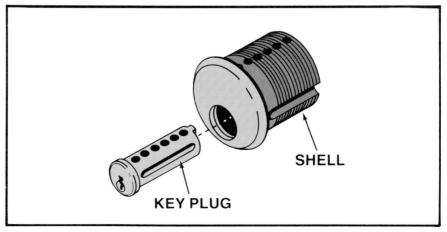

Figure 5.27 The heart of any lock or locking mechanism is the key cylinder, which is composed of an outer shell and an inner plug.

within the shell. This, in turn, rotates a tailpiece or cam at the rear of the plug to move the latch or bolt within the lock assembly.

Better-quality cylinders are made with components that resist picking and drilling. New designs incorporate such features as interlocking tumblers, which make locks virtually pick-proof. To prevent the cylinders from being drilled, case-hardened steel rods and shields (Figure 5.28) are built into the front of the cylinder.

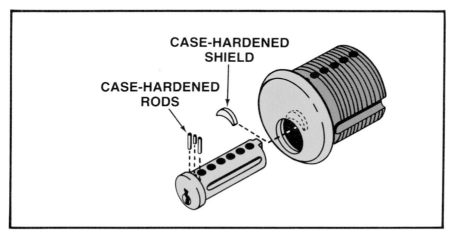

Figure 5.28 Drill-resistant lock cylinders contain case-hardened steel rods and shields that effectively stop a drill bit from reaching the tumblers.

ELECTRIC LOCK

Electric locks on entrance doors are found on many types of occupancies, most often where a high level of security is required. Such locks may be operated by inserting an encoded plastic or paper card (Figure 5.29), by encoding a combination on a numeric

Figure 5.29 One type of electric lock is activated by inserting an encoded paper or plastic card.

Figure 5.30 Another type of electric lock is activated when the correct number combination is entered by pushing buttons on a numeric key pad.

Figure 5.31 An auxiliary lock is any lock that is installed in addition to the original lock to increase the security of a door. It may be an auxiliary deadbolt, rim lock, or even a padlock.

input device (Figure 5.30), or by activating a fingerprint scanner. Many locks are controlled by a person located in a remote location, as when a condominium resident communicates over an intercom to a person seeking entrance. The resident can unlock the main entrance door by activating a switch within his or her dwelling that unlocks the electric lock on the entrance door, allowing the visitor to enter the building.

When an electric lock is activated, the strike is retracted from the latching device, which is usually a common latch bolt. It is operated by a 10- or 12-volt electric solenoid that causes it to move back into the frame far enough to clear the stationary latch bolt, allowing the door to open.

AUXILIARY LOCKS

The word "auxiliary" is a general term used to describe any lock added to a door to provide more security than the original installation offers. An auxiliary lock is usually installed on or in the door above an existing lock (Figure 5.31) and should be approached from the standpoint that it is the more substantial lock of the two. This means that it will provide more resistance to entry than the lock below. As a general rule, as locks are progressively added to a door, each is more resistant to forcing than the previously installed lock. Two of the most common auxiliary locks are the rim lock and the tubular deadbolt, previously described.

Padlocks can also be called auxiliary, although they are often the single means of securing a door or gate. Devices such as drop bars and surface bolts are also classified as auxiliary locks.

Padlock

Modern padlocks come in a variety of shapes and sizes, and vary in quality of materials and construction. Despite design differences, padlocks secure in the same basic way — by joining two separate units into one inseparable unit. This may occur when two links of chain are padlocked together, as when a gate is secured, or when a door hasp is closed over its loop and padlocked.

Padlocks are composed of the same basic components (Figure 5.32) although they may differ from conventional shape because of lock design. The main padlock body contains a locking mechanism, which is usually of the tumbler or disk type. The locking mechanism engages the shackle on one end or at both ends. This is significant from a forcible entry standpoint because a padlock that locks only one leg of the shackle requires only a single cut to disable the lock. When the locking mechanism acts on both legs of the shackle, two cuts must be made to break security (Figure 5.33).

Better quality padlocks have a number of features that make them more resistant to forcing than locks of inferior quality.

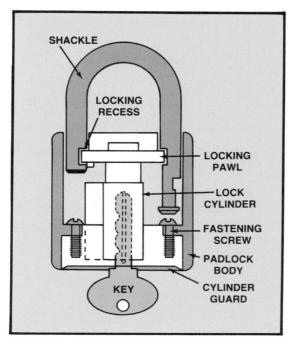

SHACKLE

LOCKING
RECESS

LOCKING
PAWL

LOCK
CYLINDER

FASTENING
SCREW

PADLOCK
BODY

CYLINDER
GUARD

KEY

Figure 5.32 All padlocks have the same basic components.

Figure 5.33 Some padlocks require that two cuts be made in the shackle because the locking mechanism engages both ends of the shackle.

Shackles made of case-hardened steel are extremely resistant to cutting with such tools as bolt cutters and hacksaws. Locks with case-hardened shackles are usually marked to identify this feature (Figure 5.34), partly as a deterrent to persons who would use illegal force against the lock. To prevent drilling out tumbler pins, padlocks may be equipped with anti-drilling rods and shields. These components are also made of case-hardened steel, a material that makes them impenetrable to most types of drill bits.

To prevent removal of the entire lock cylinder with a dent puller (see Chapter 6), better quality locks have lock cylinders made of stainless or case-hardened steel, often anchored behind a rim or guard that prevents their removal. Newer designs of lock cylinders are virtually pick-proof due to such features as interlocking pins and built-in anti-picking devices.

Figure 5.34 A case-hardened padlock shackle is usually stamped "case hard" or "hardened."

Answers on page 263

Answer each of the following questions in a few words or short phrases:

1. List the five categories of locks.

    A. _____

    B. _____

    C. _____

    D. _____

    E. _____

2. Why are doors with key-in-knob locks usually so vulnerable to prying?

    _____

    _____

    _____

3. What are the three basic configurations of rim locks, in order of resistance to prying (least resistant first)?

    A. _____

    B. _____

    C. _____

4. What are the four basic configurations of exit devices? (Indicate whether they are found on single swinging [SS] doors or on double swinging [DS] doors.)

	SS	DS
A. _____	☐	☐
B. _____	☐	☐
C. _____	☐	☐
D. _____	☐	☐

5. When two locks are encountered on a door, which is usually the stronger of the two, the lower or the higher? Why?

    _____

    _____

    _____

    _____

6. What are the two most common types of auxiliary locks?

    A. _____

    B. _____

Determine whether the following statements are true or false. If false, state why:

7. Locks that look similar from the outside of a door do not necessarily operate in a similar manner.

   ☐ T   ☐ F   _____

   _____

8. Auxiliary deadbolts are difficult to force by prying but are relatively easy to cut.

   ☐ T   ☐ F   _____

   _____

9. The latch on an electric lock is usually a common latch bolt.

   ☐ T   ☐ F   _____

   _____

Select the choice that best completes the sentence or answers the question:

10. Forcing methods for a locking assembly are based on
    A. whether the lock is light, standard, or heavy duty
    B. how the lock is mounted on the door and frame
    C. the internal mechanism of the lock
    D. the length of the latch bolt

11. A deadlocking latch bolt is fitted against the latch bolt to prevent it from being retracted by a tool on some
    A. cylindrical locks
    B. mortise locks
    C. rim locks
    D. all of the above

12. Deadbolts are found on all lock categories *except*
    A. bored
    B. mortise
    C. rim
    D. pre-assembled

13. "Key-in-knob" locks are
    A. bored locks
    B. mortise locks
    C. pre-assembled locks
    D. any of the above

14. Exit devices are designed to operate by
A. pulling
B. pushing
C. turning
D. sliding

Match the lock in the left column with the correct description in the right column:

Category	Description
____ **15.** Cylindrical	A. Actuated by a bar, mounted horizontally, operates by pushing
____ **16.** Mortise	B. Installed in two round holes bored at right angles in a door
____ **17.** Rim	C. Has a locking mechanism in the knob with a metal facepiece behind the knob
____ **18.** Pre-assembled	D. Has a threaded lock cylinder anchored by a set screw
____ **19.** Exit device	E. Surface mounted, with a cylinder recessed into the door

Forcing
Locks

6

NFPA STANDARD 1001
FORCIBLE ENTRY
Fire Fighter I

3-2 Forcible Entry

3-2.1   The fire fighter shall identify and demonstrate the use of each type of manual forcible entry tool.

Fire Fighter II

4-2 Forcible Entry

4-2.1   The fire fighter shall identify materials and construction features of doors, windows, roofs, floors, and vertical barriers and shall define the dangers associated with each in an emergency situation.

4-2.2   The fire fighter shall identify the method and technique of forcible entry through any door, window, ceiling, roof, floor, or vertical barrier.

Chapter 6
Forcing Locks

Firefighters who have a thorough understanding of how locking devices work and the methods by which locks can be forced are better prepared to make successful forcible entry than those who have not. Locking devices offer the firefighter one of the best means of gaining entrance to a building in the least amount of time and with the least amount of damage. The degree of damage varies depending upon the method used, which in turn varies with the type and quality of lock. Some of the methods described in this chapter cause little or no damage to lock components. Others cause complete destruction of the lock, but leave the door and frame unharmed.

During the size-up process, it is important to anticipate the relative amount of damage occurring with through-the-lock entry as opposed to forcing the door and frame components. For example, it may be less damaging to break a pane of glass and reach through to unlock a door than to destroy the lock and retract its deadbolt. Discussions of the relative costs of doors, frames, glass, locks, and other components should be a part of forcible entry training. Knowing the labor costs for repair and replacement of doors, windows, glass, and locks gives firefighters a better idea of the potential costs of damages caused by forcible entry.

The amount of damage caused during forcible entry operations does not necessarily increase when rapid entry is required, as to a burning building. Well-trained personnel should be able to make entry with no more damage than would occur when using the same method under less urgent conditions. The key to this ability, however, is in *preparation*. Firefighters must be trained in not only forcible entry techniques, but also knowledgeable about building features that allow the most expedient entry with the least amount of damage. This requires careful pre-incident planning.

Through-the-lock entry requires only a few tools, most of which are relatively inexpensive. Some are small enough to carry in a coat pocket. They include the K-tool, the "A"-type lock puller, key tools, vise grips, dent pullers, and shove knives (Figure 6.1). More customized tools may also be used as new methods are learned and more refined techniques are developed. Customized tools are often designed and made in the fire department shop.

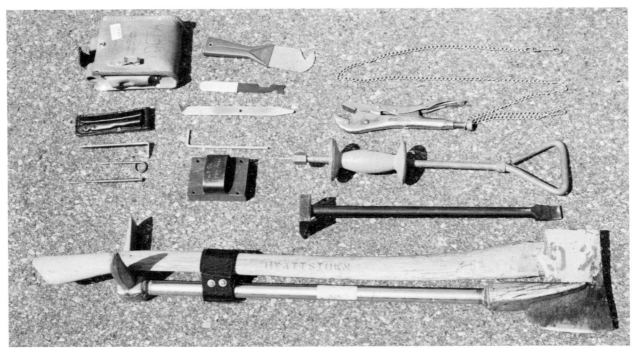

Figure 6.1 Through-the-lock entry tools.

FORCING BORED (CYLINDRICAL) LOCKS

Bored locks include key-in-knob locks and auxiliary deadbolts. Both types are installed in the same manner within holes that have been bored into the door face and edge.

Key-in-Knob Locks

Key-in-knob locks are vulnerable to several methods of through-the-lock entry. This type of lock has an outside doorknob with a key-activated lock cylinder in the center, while the inside knob usually contains a push or turn button that activates the locking mechanism. Two screws join the knobs and hold the unit in place within the bored opening. A beveled spring latch slides inside a tube in the edge of the door and extends 1/2 to 3/4 inch (13 mm to 19 mm) into the strike. The short latch allows some doors to be pried far enough from the frame to allow the latch to slip free so the door can be opened. More solid doors set in sturdy frames are not as easy to pry. In this case, it is usually appropriate to either retract the latch or to remove the lock.

RETRACTING THE LATCH (LOIDING)

This method works best on out-swinging doors. The technique is often referred to as "loiding," from the old practice of slipping a piece of celluloid (credit card) between the door and frame to push back the latch (Figure 6.2). A linoleum knife or a shove knife is a good tool for this type of operation. Insert the knife between the door and the frame over the beveled spring latch. Because the bevel is on the inside of the latch face, simultaneously push down and pull back against the bevel to retract the latch from the strike (Figure 6.3). As the latch clears the strike, pull the door open. This process does not work on a deadlocking latch, which has an adjacent pin that locks the latch into place when external force is used against the beveled end.

The "loiding" technique works on in-swinging doors only if the stop on the frame can be pried free to allow insertion of the shove knife. In this case, the latch bevel faces the outside, so push the end of the knife blade against the bevel face to retract the latch from the strike. Push the door inward as the latch clears the strike.

Figure 6.2 "Loiding" a door latch.

Figure 6.3 To loid an out-swinging door with a shove knife, pull the knife down and outward against the latch bevel. Pull the door open as the latch clears the strike.

REMOVING THE KNOB

Another way to retract the latch from the strike is by activating the internal mechanism from within the bored opening of the door, which is located directly behind the doorknob. An advantage of this technique is that, in locks of the deadlocking latch type, the deadlatch pin cannot prevent retraction of the latch. This method requires removing the doorknob (complete with lock cylinder) and its tailpiece, which extends beween the two knobs. This can be accomplished in several ways. The fastest way (but the most damaging to the lock), is to strike downward sharply on the knob with a sledge, flat-head axe, or other striking tool (Figure 6.4). Another way is to drive the blades of an A-tool behind the knob and pry the knob from the door (Figure 6.5). The knob should break loose from its assembly with tailpiece attached.

Removing the knob and its tailpiece exposes the internal mechanism, which usually features a small slot or half-moon-shaped opening (Figure 6.6). The tailpiece rotates through this opening to retract the latch when the knob is turned. Insert a key tool or screwdriver into the opening and either turn or push back the mechanism to retract the latch (Figure 6.7).

Figure 6.4 To gain access to the internal locking mechanism, remove the doorknob by striking it downward sharply with a striking tool.

Figure 6.5 Another method to remove a doorknob is by prying it loose with an A-tool.

Figure 6.6 Removing the doorknob exposes the internal mechanism, which features a small slot or half-moon-shaped opening.

Figure 6.7 Insert a key tool or screwdriver into the opening and rotate or push back the tool to retract the latch.

Auxiliary (Tubular) Deadbolts

Auxiliary deadbolts contain a deadbolt of up to 1 inch (25 mm) in length. A door equipped with this type of lock cannot easily be pried far enough apart to allow the deadbolt to clear the strike. For this reason, through-the-lock entry is especially effective.

Auxiliary deadbolts can sometimes be pulled apart to provide access to the operating mechanism. Locks of standard construction consist of a large trim ring that contains a locking cylinder on the outside of the door, a tailpiece or stem that extends from the rear of the cylinder through a slot, cross, half-moon, or "X" opening in the deadbolt assembly, and a cylinder or thumbturn on the inside of the door. Two screws, usually of brass, hold the entire assembly together by joining the outer cylinder to the inner cylinder or thumbturn plate. The screw ends are set into posts that may be of white metal casting.

To force an auxiliary deadbolt lock, drive the fork or adz of a Halligan tool down behind the trim ring and pry the cylinder from the door (Figure 6.8). The screw posts or brass screws will break, allowing the ring and cylinder to fall free. The deadbolt operating mechanism will then be exposed, allowing a view of the tailpiece opening. Insert the stem end of the key tool or a screwdriver into the opening and rotate it to retract the bolt.

Figure 6.8 To force an auxiliary deadbolt lock, drive the adz end of a Halligan tool down behind the trim ring and pry the cylinder from the door.

The K-tool, designed primarily for pulling rim and mortise lock cylinders, is occasionally useful for pulling the cylinder of an auxiliary deadbolt. Most auxiliary deadbolt cylinders are too large to fit into the working area of the K-tool. Cheaply made cylinders, however, collapse inward as pressure is applied to slide the sharp K-shaped edges over the metal body.

Newer, high-security auxiliary deadbolts have high alloy steel trim rings with internal collars around the lock cylinder on the outside of the door. Internal components are of the same heavy-duty quality. Screws of up to ¼-inch (6 mm) made of high tensile steel hold the lock components together. The deadbolt may be more than 1 inch (25 mm) long and have a hardened pin inside that is highly resistant to sawing. It is extremely difficult to pull apart these high-security deadbolts using a prying technique. Prying operations on these locks cause significant damage to the lock and possibly the door. Choose a more vulnerable site for entry, if possible. If the door and frame are made of wood, the door can probably be forced more easily than the lock.

REMOVING CYLINDER GUARDS

Before discussing how to force mortise and rim locks, which involves removing a lock cylinder, it should be noted that in high-crime areas these locks are often covered with cylinder guards. A typical unit consists of a rectangular steel plate that hides the entire cylinder, except for a small opening over the keyway (Figure 6.9). It is held in place by bolts at each corner of the plate mounted from inside the door. The plate may be indented on the inside surface to accommodate cylinders that are not flush mounted. It is necessary to remove the guard before pulling the cylinder.

The guard can be removed in several ways. One of the best methods is to drive the blade of a cutting tool, such as the adz of a Halligan tool or the blade of a flat-head axe, behind the guard over each bolt (Figure 6.10). This action shears the bolts and releases the guard. Only three bolts need to be cut if the cylinder is flush mounted. The fourth bolt can remain in place to act as a

Figure 6.9 A cylinder guard is used in high-crime areas to prevent cylinder removal.

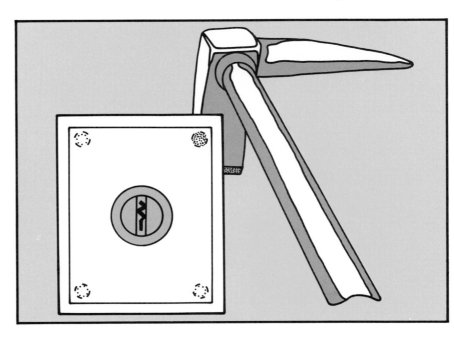

pivot so the guard can be rotated out of the way. If the cylinder protrudes from the door, cut all four bolts to completely remove the guard.

FORCING MORTISE LOCKS

The mortise lock cylinder is the most vulnerable part of the mortise lock. The best way to force a mortise lock is to remove the cylinder to gain access to the internal mechanism. Once the cylinder is removed, the internal mechanism can be manipulated to retract the deadbolt and/or latch.

Removing the Mortise Lock Cylinder

There are two ways to remove the cylinder: unscrewing and prying. Unscrewing is more time consuming than prying because the cylinder is fine threaded, but it causes the least amount of damage. Neither method can be used if the cylinder is flush mounted.

The mortise lock cylinder is threaded directly into the lock casing. A set screw is used to fix the unit in place. A small groove in the side of the cylinder accepts the screw, which may be made of brass or of a much harder metal, depending upon the lock quality. The set screw is the only resistance to removal by unscrewing.

UNSCREWING THE CYLINDER

A tool specifically designed for unscrewing a lock cylinder is the cylinder removal tool used by locksmiths (Figure 6.11). Its only drawback is that it does not give the leverage for breaking the set screw that vise grips provide.

Vise grip pliers (the 9-inch [229 mm] size with curved jaws works well) may also be used to remove a lock cylinder. After adjusting the jaws for the diameter of the cylinder, clamp the pliers to the cylinder from the side to provide as much contact with metal as possible. With the pliers, turn the cylinder one-eighth of a turn clockwise to break off the tip of the set screw. Then turn the cylinder counterclockwise to unscrew it from the lock casing (Figure 6.12). If the set screw is made of a hard metal, the tip may not

Figure 6.12 Clamp vise grip pliers to the cylinder, rotate the cylinder one-eighth turn clockwise to break the tip off of the set screw (above), then rotate the cylinder counterclockwise to unscrew it from the door (below).

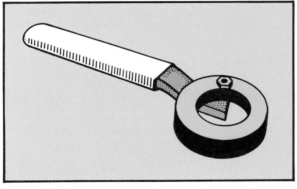

Figure 6.11 A cylinder removal tool, used by locksmiths.

break off. In this case, the tip only bites into the softer cylinder body. When this happens, the set screw scores a deep groove around the cylinder as the cylinder is rotated, which prevents further retraction of the cylinder from the casing. If this occurs, the cylinder should be pulled.

PULLING THE CYLINDER

A mortise lock cylinder can be pulled using any one of several tools. The amount of resistance to this operation is based on several factors:

- Quality of the cylinder and lock casing
- Type of metal in the cylinder and casing
- Cylinder protection (guard, collar)
- Accessibility (close to floor; recessed)

The most significant problem with pulling a mortise lock cylinder is that very little of the cylinder body protrudes from the door. This makes it difficult to use a standard prying tool, such as a Halligan tool, because the prying end cannot be easily driven behind a trim ring to provide enough resistance for leverage. If pressure is applied to the trim ring only, it usually collapses and the cylinder remains in place. Most tools cannot get a sufficient "bite" into the side of the cylinder wall to remain solidly attached to the body during the entire prying operation.

The K-tool is designed specifically to pull rim and mortise cylinders. It has two keenly sharpened blades of hardened steel that "bite" into the cylinder walls when applied. One blade is straight and the other is V-shaped, forming a "K"-shaped opening that slides over the cylinder (Figure 6.13). A loop of steel on the outside of the tool accepts the adz end of a Halligan-type bar.

To apply the K-tool, hold it in one hand with the opening of the tool in a position to slide the blades onto the lock cylinder (Figure 6.14). With a flat-head axe or Halligan tool, strike the K-tool with light blows, driving it over the cylinder so that the blades go behind the trim ring and face of the cylinder (Figure 6.15). When the blades are driven in deeply enough that the tool does not need to be held in place by hand, insert the adz end of a Halligan tool into the K-tool loop. Using light blows of the flat-head axe against the flat portion of the adz, continue to drive the blades of the K-tool behind the face of the lock cylinder until they bite into the main body (Figure 6.16). Remember that the blades of the K-tool are hardened steel and that lock cylinders are usually formed from a soft metal, like brass, and that driving the tool too far into the cylinder body could cut it in half. Apply leverage to the K-tool through the Halligan tool handle until the cylinder threads strip loose from the casing. This may occur quite suddenly, so be prepared for a sudden release of resistance.

Figure 6.13 The K-tool jaws are designed to bite into a cylinder.

Figure 6.14 Position the K-tool over the cylinder trim ring.

Figure 6.15 Tap the K-tool into position behind the cylinder trim ring.

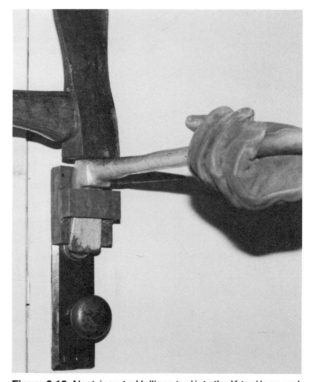

Figure 6.16 Next, insert a Halligan tool into the K-tool loop and drive the K-tool further onto the cylinder with a flat-head axe.

The A-tool can also be used to pull mortise and rim cylinders. It features an "A"-shaped set of case-hardened steel blades that will bite into and grip the main body of the lock cylinder. Its handle provides a built-in means of applying leverage force to the blades.

Using the A-tool is similar to using its counterpart, the K-tool. With a separate striking tool, drive the blades into place behind the trim ring until they bite into the cylinder wall deeply enough to provide a solid grip (Figure 6.17). Then apply leverage to the tool through its handle until the cylinder threads strip loose from the casing. The cylinder may break loose suddenly, so be prepared for a sudden release of resistance.

Figure 6.17 Drive the blades of an A-tool into position behind the trim ring of a lock cylinder so that they bite into the cylinder wall.

Retracting the Mortise Lock Deadbolt and Latch

The presence of a *threaded* cylinder in a lock indicates one of three types of *mortise* locks: deadbolt, deadbolt and latch, or pivoting deadbolt. Before removing the cylinder, note the location of the keyway. The keyway is in a plug that rotates inside the lock cylinder when the key is turned. It is set to one side of the cylinder face. Imagine a clock dial on the cylinder face. To establish a relationship between the location of the keyway and the locking mechanism behind it, imagine the keyway position at 6 o'clock (Figure 6.18). A cam at the rear of the keyway turns to activate the locking mechanism, moving it between the 5 and 7 o'clock positions (Figure 6.19).

When the cylinder is removed, insert the right-angled portion of a key tool through the vacant opening and rotate the tool away from the deadbolt strike, moving the locking mechanism between the 5 and 7 o'clock positions (Figure 6.20). This retracts the deadbolt from its strike and frees the door to swing open.

If there is a spring lock on the deadbolt mechanism, use the key tool to depress the spring and then move the mechanism between 5 and 7 o'clock. The lock may also contain a spring-operated lever that controls a locking doorknob. The lever in this kind of arrangement is between the 3 and 5 o'clock positions or between the 7 and 9 o'clock positions. After retracting the deadbolt, turn the key tool sideways, depress the lever, and rotate the tool to retract the latch.

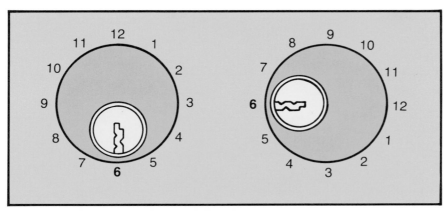

Figure 6.18 As a reference, imagine the lock cylinder keyway to be positioned at 6 o'clock. Two views are shown above — keyway at the bottom, keyway at the side.

Figure 6.19 A cam at the rear of the keyway moves the locking mechanism between 5 and 7 o'clock.

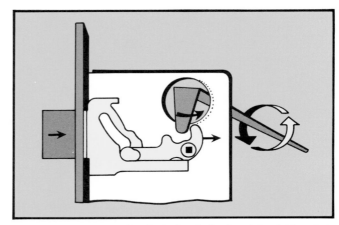

Figure 6.20 After removing the lock cylinder, insert the right angled end of the key tool and move the locking mechanism between 5 and 7 o'clock to retract the deadbolt.

Another type of mortise lock is the pivoting deadbolt, commonly found in modern aluminum-frame glass doors. It is opened the same way as other mortise deadbolt locks. Note the keyway position in relation to the position of the cylinder within the stile (Figure 6.21). Once the cylinder is pulled, insert a key tool to de-

Figure 6.21 Before pulling the cylinder, note the position of the keyway within the cylinder.

press the spring-loaded mechanism that activates the pivoting deadbolt (Figure 6.22). Rotate the key tool to move the mechanism between the 5 and 7 o'clock positions (Figure 6.23). This will cause the deadbolt to pivot out of the strike and back into the lock case.

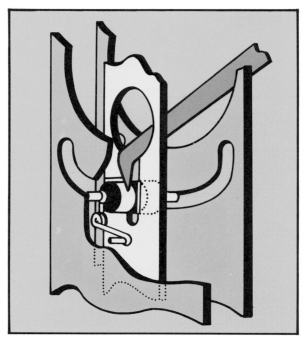

Figure 6.22 After pulling the cylinder from a pivoting deadbolt lock, insert a key tool to depress the spring-loaded mechanism.

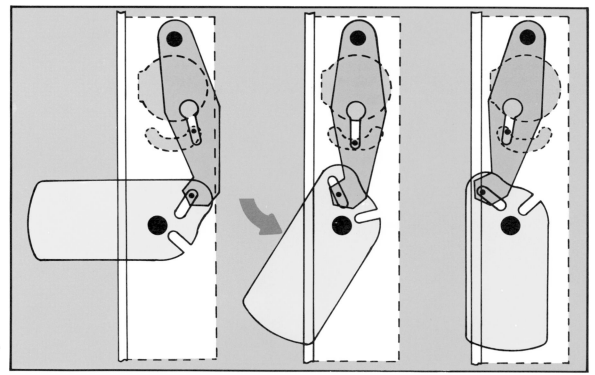

Figure 6.23 Rotate the key tool to move the mechanism between the 5 and 7 o'clock position. This will cause the deadbolt to pivot 90 degrees out of its strike.

FORCING RIM LOCKS

A rim cylinder appears identical to a mortise cylinder when it is mounted in a door. A rim cylinder, however, is anchored to the door by two screws that protrude from a back plate on the inside of the door, while a mortise cylinder is threaded into place within a lock casing.

The best way to recognize the difference between the two types of cylinders is by noting their relative position in the door. A mortise cylinder is usually set at the center edge of the door, near the knob or handle. A rim cylinder is usually located higher on the door because it is part of an auxiliary lock, such as an interlocking deadbolt. This means that it was added to the door to supplement the original lock. In the case of narrow-stiled glass/aluminum doors, a mortise cylinder (pivoting deadbolt) is used because there is not enough room on the stile to accommodate a rim lock.

A rim cylinder can be removed by either twisting it off the door or by prying it away from the door. To twist the cylinder free, clamp a pair of vise grip pliers to the cylinder (in the same way as described for the mortise cylinder) and turn the cylinder in either direction to break the mounting screws. A K-tool or A-tool can be used to pry the cylinder out of the door. Again, set either tool into place behind the trim ring (Figure 6.24) and pry the unit from the door by snapping off the screws (Figure 6.25). Once it is removed,

Figure 6.24 Drive a K-tool (left) or an A-tool (right) down behind the trim ring of the rim cylinder.

Figure 6.25 Pry the cylinder from the door by levering the tool upward, breaking the mounting screws.

inspecting the cylinder reveals what type of lock it is. A rim cylinder is not threaded and has a tailpiece protruding from the rear (Figure 6.26). A mortise cylinder has a threaded case and a cam at the rear.

Regardless of how badly damaged a rim cylinder becomes during the prying operation, the locking mechanism inside the door is not usually damaged. As long as a slot is visible through the empty cylinder hole, a key tool or screwdriver can be inserted into the slot and turned to open the lock as easily as using a key, with one important exception: if the key tool can be inserted but will not turn in either direction, the lock may have been latched or bolted from the inside with a night latch. In this case, suspect that someone is inside and base further action on this possibility (search the structure if it is charged with smoke).

When a key tool will not work to retract the locking mechanism, the best way to complete through-the-lock entry is to punch the inside portion of the lock from the door. To do this, use the pike end of a Halligan tool and a flat-head axe or sledge. Insert the pike into the cylinder hole against the back of the lock (Figure 6.27). Strike the back of the pike sharply with the striking tool. This will tear the lock loose by stripping the mounting screws that hold it to the door.

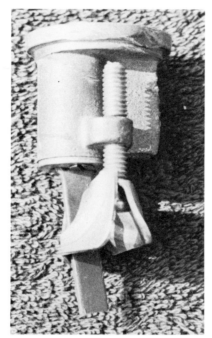

Figure 6.26 A rim cylinder has a stem, or tailpiece, protruding from the rear of the cylinder.

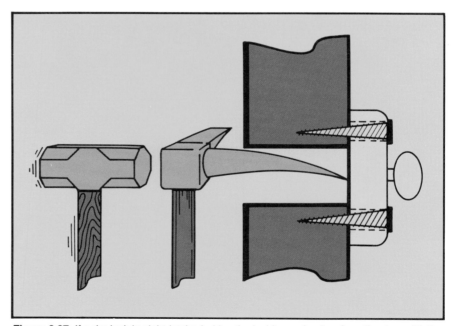

Figure 6.27 If a rim lock is night-latched, drive the inside mechanism from the door with the pike end of a Halligan tool.

Some rim locks have a spring-loaded security shutter that snaps across the opening when the tailpiece is pulled from the slot. The shutter must be retracted before a screwdriver or key tool can be inserted to retract the deadbolt or latch. The best tool for this job is a tool that closely resembles a sharpened ice pick.

Place the tip against the end of the shutter and with steady pressure push back the shutter until the slot becomes visible (Figure 6.28). Then insert the key tool or screwdriver into the slot and retract the deadbolt or latch (Figure 6.29).

Improved shutter designs feature a tab that drops down when the shutter is closed to lock it in place. In this case, an additional tool is required to open the shutter. Use a small tool such as a curved pick to lift the shutter plate, then slide the shutter open (Figure 6.30). With the shutter retracted, insert a key tool into the slot and retract the deadbolt or latch.

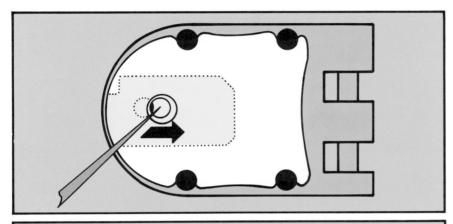

Figure 6.28 Retract the shutter of a rim lock with a sharp pick tool.

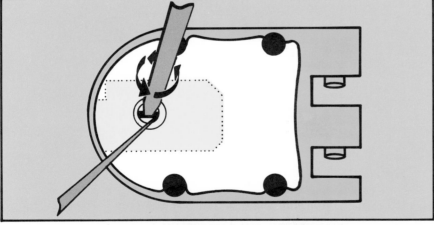

Figure 6.29 Holding the shutter back, insert the key tool and rotate the tool to retract the deadbolt or latch.

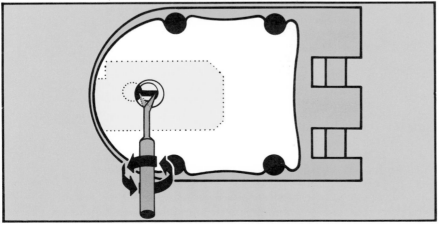

Figure 6.30 If the shutter has a retaining tab, insert a curved pick tool and find the opening. Simultaneously lift and retract the shutter plate. Then insert the key tool and retract the deadbolt or latch.

When smoke conditions prevent an unobstructed view of the locking mechanism, or when there is simply no time to retract shutters and deadbolts, punch the inside lock free with a Halligan tool.

One type of interlocking deadbolt lock features a security plate on the outside of the door. The lock cylinder is attached directly to the plate, and the plate is mounted to the door with bolts at each corner from the inside. This prevents the cylinder from being pulled with a K-tool or A-tool. To pull the cylinder of this type of deadbolt, the entire assembly must be removed. This can be done by shearing the bolts with the adz of a Halligan tool driven behind the plate at each corner. Once the bolts are sheared, the plate, cylinder, and tailpiece can be removed to reveal the back of the locking mechanism on the inside of the door. Then insert a key tool or screwdriver to retract the deadbolt.

FORCING PREASSEMBLED (UNIT) LOCKS

There are some high-security key-in-the-knob lock sets that do not respond to most through-the-lock methods. These preassembled locks are some of the most force-resistant locks available.

The best way to force this type of lock is through the knob, either by breaking it or prying it from the assembly. Break the knob from the door by striking it sharply downward with a heavy striking tool. Expect resistance to this action from higher quality unit locks. Locks made of heavy-duty materials are assembled in such a way that it is difficult to break them off without doing damage to the door. Prying is even more difficult and should only be attempted with heavy long-handled tools, perhaps with an additional fulcrum. Expect to do some damage to the door when prying these locks.

With either method, after removing the knob, use a key tool or screwdriver to operate the internal locking mechanism. This retracts the latch and frees the door to swing open. The deadlocking pin does not interfere with retracting the latch.

In view of the heavy-duty nature of the preassembled lock, consider measures other than through-the-lock entry. If door and frame construction allow, use pry tools to force them apart, clearing the latch from the strike. Also consider cutting the door in half or cutting the lock out of the door. Cutting methods can still be considered, but only as a last resort because of the damage they cause.

FORCING EXIT DEVICES

Exit devices are used on exit doors in places of public assembly. They are equipped with panic hardware and come in rim and mortise designs. Exit devices may have rods as part of the assembly. Rim-type exit devices are most common.

Rim lock mechanisms are mounted on the inside surface of a door. They may have a spring latch bolt that extends into a strike on the frame or mullion, or vertical rods that extend to the top and bottom of the door to push against spring latches. The latches engage strikes on the frame and in the floor. Mortise lock mechanisms are mounted inside the body of the door. They may also have a spring latch bolt that extends into a strike in the frame or mullion, or vertical rods inside the door to push against top and bottom spring latches that engage strikes built into the frame and floor.

One of the simplest methods of forcing an exit device is by hooking the crossbar with a piece of wire and pulling back to activate the panic hardware. A short piece of stiff wire, hook-shaped on one end, will work very well for this operation (Figure 6.31). Insert the hook-shaped end of the wire between the door and frame or mullion, hook the crossbar, and pull outward to retract the latch or rods that secure the door.

For both types of installations, a lock cylinder is used to secure the door from the outside. The lock cylinder is like other rim or mortise cylinders and can be removed by the methods described previously. Once the cylinder is removed, use a key tool or screwdriver to retract the latches.

If the cylinder is mounted in an aluminum-frame glass door and cannot be pulled without damaging the door frame and panic lock mechanism, break the glass and reach inside to trip the panic hardware. If the panic hardware is chained, use bolt cutters to cut the chain to allow the door to swing free (Figure 6.32).

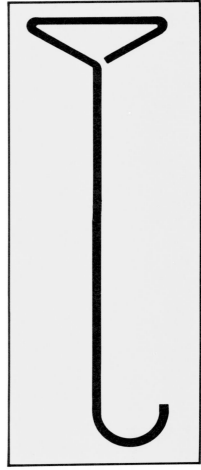

Figure 6.31 A short piece of stiff wire with a hooked end is a handy tool for forcing an exit device.

Figure 6.32 Some building owners chain panic hardware on double doors when the building is locked up after hours. Cut the chain with bolt cutters.

DRILLING LOCK CYLINDERS

One of the simplest ways to go through a lock is with a drill. Successful drilling requires, however, that the lock cylinder be of the tumbler type. A tumbler lock operates when a key is inserted into the keyway, which raises a set of pins so that a shear line is created within the plug (Figure 6.33). At this point the key can be turned because no pins obstruct the plug's rotation.

Drilling, in effect, creates a new shear line. To accomplish this, drill a very small hole (about the same diameter as a standard tumbler pin — .115 inch [2.9 mm]) into the face of the cylinder at the height of the shear line (Figure 6.34). If the hole is correctly aligned, the section of pins that obstruct the shear line are drilled away, creating a new shear line. Carefully remove the bit, insert a small screwdriver into the keyway, and turn it to activate the locking mechanism (Figure 6.35).

Drilling works on any tumbler-type lock, whether it is in a rim or mortise cylinder, padlock, or key-in-knob lock. However, it does not work on locks that have been reinforced to prevent this type of action. Such locks contain shields of case-hardened steel located immediately behind the cylinder face at the shear line that block drilling. Other locks also have plug faces of high carbon steel that are virtually impenetrable to most types of drill bits.

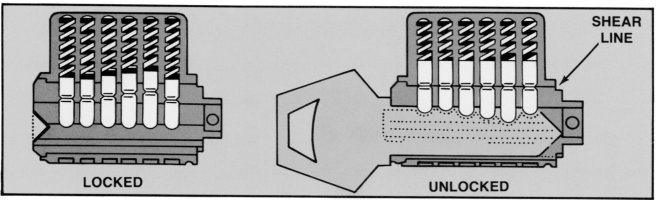

Figure 6.33 When a key is inserted into the keyway of a tumbler lock, a set of pins (tumblers) are aligned to create a shear line. This allows the plug to rotate within the shell.

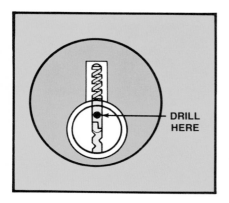  

Figure 6.34. Drill a key plug just above the keyway opening to create a new shear line. Center punch a pilot hole to prevent the drill bit from drifting. Drill straight into the cylinder.

Figure 6.35 Use a small screwdriver to rotate the plug and retract the locking mechanism.

FORCING PADLOCKS

Padlocks provide security for a wide range of doors, gates, and other entry points to property. They come in many designs, sizes, and vary from lightweight to extremely heavyweight construction. No matter what the type, however, all padlocks are relatively vulnerable to forcing because of one major reason: accessibility. A variety of tools and methods can be used to defeat virtually every type of padlock, no matter how well made, because the entire lock is exposed. The following are basic approaches to forcing padlocks:

- Prying
- Twisting
- Cutting
- Spreading
- Drilling
- Striking

Prying can be attempted when the mechanism to which the padlock is attached, such as a door hasp, is mounted externally. In this case, insert an appropriate prying tool under the mechanism and pry it loose from the surface (Figure 6.36).

Twisting is probably one of the oldest methods used for forcing padlocks. The basic premise is that if enough twisting force is applied with a large enough tool, something breaks. This is usually true, especially if the padlock is attached to a relatively rigid device such as a door hasp. This method does not work as well when the lock is attached to a chain or a cable. Twisting involves placing the fork of a Halligan tool over the shackle and rotating

Figure 6.36 Some hasps are relatively easy to pry from a door. Insert the adz of a prying tool under the hasp and pry outward.

the handle in either direction until the weakest component breaks (Figure 6.37). Be aware, however, that if the hasp is twisted out of shape, the lock components can bind, thus preventing the assembly from opening. Another method involves placing the fork of a Halligan tool over the shackle and, using a fulcrum, prying the entire assembly from its anchor point.

An effective, but destructive, method of forcing a padlock is by striking. As stated before, the key to effective tool use in forcible entry is to select the proper tool for the job. Make sure the striking tool is heavy enough to do the job with as few blows as possible. It is best to place the fork or adz of a Halligan tool over the padlock body, then to strike the adz to drive the body from the shackle (Figure 6.38).

Figure 6.37 Place the adz of a Halligan tool over the shackle, then twist it until the hasp or lock breaks.

Figure 6.38 To break a padlock, place the adz of a Halligan tool over the padlock hasp and strike the tool (rather than the lock) with a heavy sledge.

Cutting can be done with several types of tools. No matter how a padlock is cut, however, remember that cutting one leg of a shackle does not create an opening if the lock is of the type in which the locking mechanism acts against both legs (sometimes called "heel and toe locking"). Most better-quality locks are of this design, so two cuts must be made to open the shackle.

The bolt cutter has traditionally been one of the most common tools for cutting a padlock. However, it is fast becoming obsolete for this purpose because of the predominance of case-hardened shackles. Bolt cutter jaws are no match for the high tensile strength metal found in padlock shackles and the jaws usually dent when force is applied to 1/4-inch (6.4 mm) or larger shackles. If the word "hardened" appears on a lock, choose a method other than cutting with a bolt cutter. When making a cut on a nonhardened shackle, position the shackle deep within the jaws rather than at the extreme tips (Figure 6.39).

Figure 6.39 Position the shackle well within the bolt cutter jaws for maximum cutting effectiveness.

Another tool used for cutting is the circular saw. Use a metal cutting blade to cut the shackle or the component to which the padlock is attached (chain or hasp, whichever is softer). When cutting the shackle, attach a pair of vise grip pliers to the lock body and hold tension on the lock in such a way as to stabilize the lock in a firm cutting position. For increased safety, use a short length of chain attached to the plier handle to increase the distance between the holder's hand and the saw blade (Figure 6.40). Cut both legs of the shackle at the same time by spanning the shackle opening with the blade.

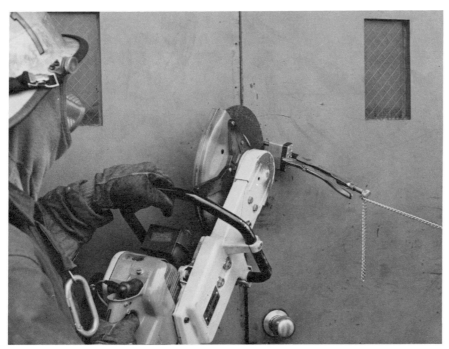

Figure 6.40 When cutting the shackle with a circular saw, a second person should hold tension on the padlock with vise grips.

If time allows, a cutting torch can also be used for cutting. Use care when operating the torch near combustible materials. Use a pair of vise grip pliers to hold the lock in place, but remember that the pliers and chain conduct heat, so wear heavy gloves. The greatest advantage in using the cutting torch is that it cuts through any type of padlock. When there is doubt as to whether other tools or methods will be effective, use a cutting torch.

One of the newer tools to be developed for forcing padlocks is the "duck-billed" lock breaker, a wedge-shaped tool designed to force the shackle from the lock body as it is pounded through the shackle opening. To use the lock breaker, insert the pointed end into the opening and strike the back of the tool head with a sledge or flat-head axe, driving it in until the lock fails (Figure 6.41). Expect failure at the side of the lock body or within the internal locking mechanism as the shackle is spread apart. A hammer-head pick may also be used as a lock breaker for smaller locks.

Drilling the lock plug works well on a padlock with pin tumblers. Stabilize the lock with vise grip pliers or a similar tool. The padlock must be maintained in a stationary position while drilling to ensure that the drill hole is made in correct alignment with the pin line.

Another way to force a padlock is to pull the lock plug from the padlock body. This can be done on some locks with a dent puller (also called a "bam-bam" or "slap hammer"). Screw a hardened sheet metal screw, which is attached to the end of the tool, into the keyway at least 3/4-inch (19 mm) (Figure 6.42). Take special care to keep the tool aligned with the screw to prevent it from breaking

Figure 6.41 Drive the duck-billed lock breaker into the shackle opening with a sledge or flat-head axe.

Figure 6.42 Screw the hardened sheet metal screw of the dent puller at least ¾-inch (19 mm) into the keyway, and maintain alignment to prevent the screw from breaking off.

off, which occurs with sideways movement. Sharply rap the sliding handle back against its rear stop a number of times until the plug pulls out of the padlock body. This exposes the inside of the lock. Insert a key tool or screwdriver into the vacant plug hole and manipulate it to unlock the internal mechanism (Figure 6.43).

One of the most difficult padlocks to force is the type that has a hidden shackle. Because it is shielded within the body of the lock, the shackle is inaccessible to conventional lock-forcing tools such as duck-billed spreaders or pry bars. Only two methods will work on this type of lock: pulling the key plug or cutting the entire lock body with a torch or circular saw.

Because there are several types of these locks on the market, it is wise to become familiar with the arrangement of the shackle within the body, as well as with the way it operates. A padlock that has a steel pin (shackle) extending from the key plug through the padlock body (Figure 6.44) can be forced by cutting. Use a circular saw to make a cut across the lock body, perpendicular to the key plug, two-thirds of the distance between the keyway and the opposite side (Figure 6.45). This cuts the shackle pin and releases the lock. Another way to force this lock is to pull the key plug with a dent puller if the plug is not protected. The steel pin shackle comes out with the plug and releases the lock.

Figure 6.43 Insert a key tool into the plug opening and manipulate the lock mechanism to unlock the padlock.

Figure 6.44 A hidden-shackle padlock is resistant to conventional padlock forcing methods.

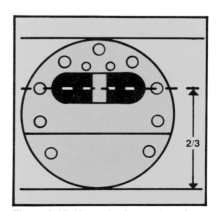

Figure 6.45 Use a circular saw to make a perpendicular cut that is two-thirds of the distance between the keyway and the opposite side.

SPECIAL PROBLEMS

There are a number of unique locks that require different treatment for through-the-lock entry. The key to success in forcing complex locks is understanding how they are assembled and how they operate. This requires locating and identifying each type of lock in the response area, then doing research to determine the unique lock characteristics. One of the best sources of information is a local locksmith, although many are reluctant to reveal trade secrets about lock weaknesses. Another source of information is the lock manufacturer, who can supply brochures and drawings of internal mechanisms.

Forcing Push-Button Locks

Push-button combination locks found on commercial buildings activate either a latch, a deadbolt, or an electric strike. Locks that activate latches or deadbolts are usually door mounted (Figure 6.46), while those that activate electric strikes are usually wall mounted.

Depending upon the design, a door-mounted unit can sometimes be pried from the door to gain access to the internal mechanism. This is possible when the mechanism is contained within a casing mounted on the door surface. Prying the casing breaks or strips out the screws that secure the casing to the front of the door. This action exposes the latch or bolt mechanism, which can then be retracted with a key tool or another appropriate tool. Another method is to drive the lock from the door with a sledge (Figure 6.47). This also exposes the latch or bolt mechanism.

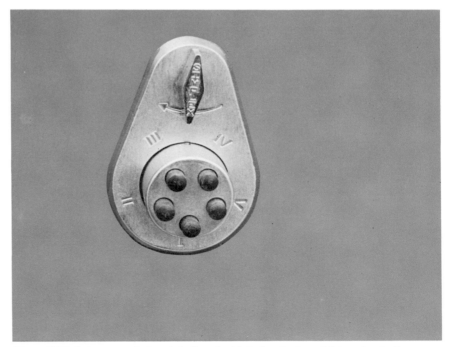

Figure 6.46 A door-mounted push-button electric lock activates a latch or deadbolt.

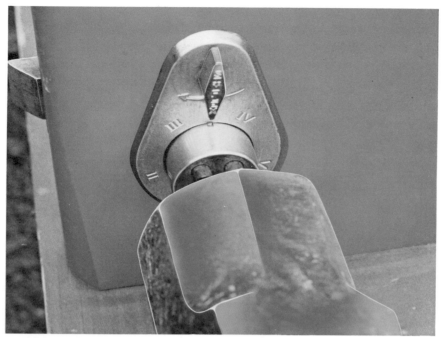

Figure 6.47 Remove a door-mounted push-button electric lock from a door by striking with a sledge; this will expose the latch or bolt mechanism.

Another common type of electronic lock uses touch sensors. It is usually very difficult to pry the sensor panel loose from the door to expose the locking mechanism. In this case, a sledge can be used to drive the entire lock from the door. Because the lock is mounted from the back of the door with screws, striking the plate with the sledge damages the entire assembly to such a degree that the screws come loose and the entire lock falls from the door.

The wall-mounted units that control electric strikes are difficult to force, primarily because of the manner in which they are assembled. It is best to consider prying the door, breaking door glass, or some other conventional method, rather than attempting to remove the lock.

Forcing Brace Locks

A popular lock in certain high-crime areas is a unique type of rim lock. This lock features a metal rod that extends from the lock to the floor (Figure 6.48 on next page). The rod acts as a brace to reinforce the door so that it cannot be battered in.

Another unique lock used in high-crime areas is the multi-bolt lock, which is mounted in the center of the door. This lock features long rods or bolts that extend in four directions from the locking mechanism into strikes in the frame and floor (Figure 6.49 on next page).

Both types of locks can be treated in the same way when forcing. Each lock uses a standard rim lock cylinder with a 3/32-inch (2.4 mm) square tailpiece. The cylinder is usually protected by a

Figure 6.48 The brace lock is a rim lock that has a metal rod extending from the lock to the floor; the rod braces the door so that it cannot be forced inward.

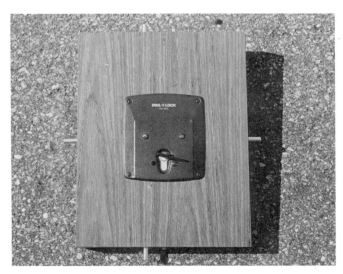

Figure 6.49 The multibolt lock features long rods that extend in all four directions from the locking mechanism into the floor and frame (a display model of one type of multibolt lock is pictured).

steel plate cylinder guard so that only the keyway shows. This guard must be removed before pulling the cylinder with a K-tool. The cylinder guard is held on the door with carriage bolts that resist prying but have relatively little shear strength. To remove this guard, drive a flat-head axe blade or Halligan tool adz down behind the guard to shear off each carriage bolt. After removing the cylinder guard, pull the cylinder with a K-tool or A-tool, insert the key tool into the square hole in the back of the lock, and rotate it to unlock the internal mechanism.

Newer versions of the multibolt lock have a geared mechanism in the door that prevents the use of a key tool. In this case, it is necessary to cut the entire locking mechanism out of the door with a circular saw and to retract the remaining exposed bolts by hand.

Forcing Barred and Bolted Doors

Sometimes doors are secured with bars or surface bolts. These devices hold the door very securely and prevent most conventional forms of forcible entry. Bars and surface bolts are often found on the rear doors of commercial occupancies and are locked from the inside by the occupant, who then leaves through another door. Since there is no external lock visible, through-the-lock forcible entry is impossible. Doors with bars are sometimes identified by bolt heads on the outside of the door (Figure 6.50), which indicates that there are brackets on the inside of the door that hold door bar(s). When bar brackets are welded into place, however, bolt heads are not visible.

Figure 6.50 Doors with bars are sometimes identified by bolt heads on the door and frame.

When a door is secured with one or more bars, the most practical forcible entry method is to cut the door with a circular saw or a chain saw. Cut the door in half from top to bottom (Figure 6.51). As the saw cuts the door, it will also cut the bar(s). Another method is to cut a small triangle in the door, then reach in and remove the bar(s) from the brackets, freeing the door to swing open (Figure 6.52).

With a door secured by a surface bolt, the methods for gaining entrance are similar to those for a barred door. Cutting the door in half will permit the bolted half-section of the door to fall free. The surface bolt is disabled at this point as it slides from its strike. Another way to force a bolted door is to cut a small triangle in the center of the door near the unhinged edge. This permits reaching inside to slip the bolt out of its strike, which frees the door to swing open.

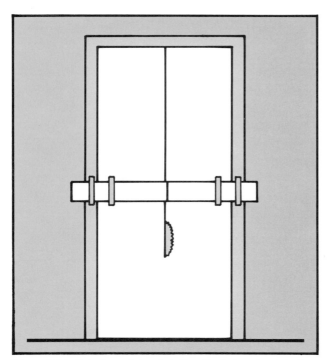

Figure 6.51 Cut a barred door in half vertically from top to bottom.

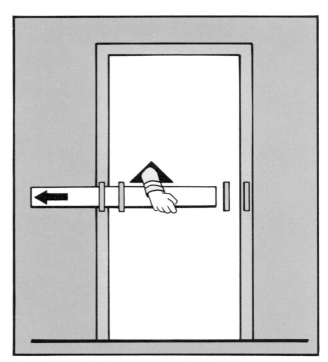

Figure 6.52 Cut a triangle in a barred door that is large enough to allow personnel to reach through and remove the bar from its brackets.

Answer each of the following questions in a few words or short phrases:

1. What is "loiding"?

   _____

   _____

2. Is loiding effective on a deadlocking latch? Why?

   _____

   _____

3. One method of forcing a bored lock is to retract the latch from the strike by activating the internal mechanism. Is this method effective on a deadlocking latch?

   _____

   _____

4. You have exposed the locking mechanism of a rim lock but cannot retract the latch with a key tool. Can through-the-lock entry be continued and, if so, how?

   _____

   _____

5. Where is a multibolt lock mounted on a door?

   _____

   _____

6. What are two recommended methods of forcing a door secured with a bar?

   A. _____

   B. _____

Complete the following statement with words or phrases that will make the statement correct:

7. A/an _____ cylinder is usually set at the center-edge of the door near the knob while a/an _____ cylinder is located higher on the door.

Determine whether the following statements are true or false. If false, state why:

8. Through-the-lock entry causes less damage than other methods of forcible entry.

   ☐ T   ☐ F   _____

   _____

Review

Answers on page 263

9. The most effective training in forcible entry includes not only practice with equipment but also classroom training on construction techniques and building features.

☐ T ☐ F _____

_____

10. A disadvantage of through-the-lock entry is that it requires expensive tools.

☐ T ☐ F _____

_____

11. The K-tool is designed primarily for pulling rim and mortise lock cylinders.

☐ T ☐ F _____

_____

12. Through-the-lock entry is especially effective on high-quality, high-security tubular deadbolts.

☐ T ☐ F _____

_____

13. Mortise locks equipped with cylinder guards are impervious to through-the-lock entry techniques.

☐ T ☐ F _____

_____

14. If the cylinder of a rim lock becomes damaged when pulled, the locking mechanism is usually not damaged.

☐ T ☐ F _____

_____

15. Twisting is an effective means of forcing a padlock attached to a chain.

☐ T ☐ F _____

_____

16. A cutting torch is effective on any type of padlock.

☐ T ☐ F _____

_____

17. A Halligan tool can be used to force open a padlock by

A. twisting ☐ T ☐ F _____

_____

B. prying ☐ T ☐ F _____

_____

C. cutting ☐ T ☐ F _____

_____

D. spreading ☐ T ☐ F _____

_____

E. drilling ☐ T ☐ F _____

_____

F. striking ☐ T ☐ F _____

_____

Select the choice that best completes the sentence or answers the question:

18. When forcible entry is accomplished by through-the-lock entry, the amount of damage varies depending upon all of the following *except* the
    A. method used to effect entry
    B. type and quality of lock encountered
    C. amount of time required to effect entry
    D. training received in through-the-lock entry

19. One method of forcing a bored key-in-knob lock is to retract the latch by activating the internal mechanism. Before this can be done the
    A. door must be pried loose from the frame
    B. doorknob must be removed
    C. cylinder guard must be removed
    D. security shutter must be retracted

20. One method of forcing a bored tubular deadbolt is to retract the latch by activating the internal mechanism. Before this can be done the
    A. door must be pried loose from the frame
    B. doorknob must be removed
    C. cylinder must be pried from the door
    D. deadbolt must be cut

21. Which is the best method of forcing a mortise lock with a flush-mounted cylinder?
    A. Prying the door
    B. Pulling the cylinder
    C. Unscrewing the cylinder
    D. Any of the above

22. A threaded cylinder on a lock indicates a
A. cylindrical lock
B. mortise lock
C. rim lock
D. unit lock

23. When a key tool cannot retract the locking mechanism of a rim lock you should suspect that
A. it was damaged when the rim cylinder was removed
B. the door is secured with a slide bolt or bar
C. there is a spring plate on a deadbolt mechanism
D. the lock was bolted from the inside with a night latch

24. What is the recommended method of forcing an interlocking deadbolt lock with a security plate on the outside of the door?
A. Shearing the bolts off the plate to expose the lock cylinder
B. Pulling the cylinder with a K-tool or A-tool
C. A, then B
D. None of the above — this lock cannot be forced

25. The most common type of exit device is the
A. rim type
B. mortise type
C. surface vertical rod type
D. concealed vertical rod type

26. Forcing a padlock by twisting would work best when the padlock is attached to
A. a chain
B. a door hasp
C. A or B
D. neither A nor B — padlocks should not be twisted

Windows
and
Window
Assemblies

7

NFPA STANDARD 1001
FORCIBLE ENTRY
Fire Fighter II

4-2 Forcible Entry

4-2.1　The fire fighter shall identify materials and construction features of doors, windows, roofs, floors, and vertical barriers and shall define the dangers associated with each in an emergency situation.

Reprinted with permission from NFPA Standard No. 1001, *Standard for Fire Fighter Professional Qualifications*. Copyright © 1981, National Fire Protection Association, Quincy, MA.

Chapter 7
Windows and Window Assemblies

Advances in technology and construction techniques have made doors and locks more secure and more resistant to forcing. Doors that are extremely difficult to force make it necessary to investigate other means of gaining entry. Forcing entry through windows is traditionally regarded as the best alternative to forcing entry through doors and, in many cases, this may be so. Buildings are constructed with windows to allow the entrance of light and sometimes to provide ventilation. Made of relatively permeable materials, windows are the least secure areas in buildings and thus prime targets for forcible entry.

Modern architectural practices, however, have vastly improved the resistance of windows to forcible entry. Current trends toward energy conservation and high security have prompted the installation of better-constructed windows. In some areas, these trends have led to an increase in the number of windowless buildings (Figure 7.1). In addition, owners are retrofitting many older buildings with windows that are more resistant to entry and have better energy conservation characteristics.

Figure 7.1 In some areas, trends toward energy conservation have led to an increase in the number of windowless buildings.

In high-crime areas, improvements in security window design mean that firefighters may find windows more difficult to force than ever before. These windows, found in both residential and commercial buildings, can pose just as formidable a barrier to entry as doors can.

Knowing how to gain entrance through windows is just as important as knowing how to gain entrance through doors. Window entry may be the only way to position hoselines to attack the seat of the fire, as well as to avoid pushing the fire toward unburned areas, which causes unnecessary damage. As with doors, efficient window entry requires the ability to gain entry both rapidly and with as little damage as possible.

This chapter describes the construction features of windows and the ways basic types of windows operate. A working knowledge of the construction and operation of typical window units is necessary to learning and using forcible entry techniques efficiently and effectively.

WINDOW CONSTRUCTION

Windows serve many functions. They allow light to the interior of a building, provide the building with ventilation, enable occupants to see outside the building, and, in some cases, allow passersby to see inside the building. Construction details of windows vary depending on their primary functions and the manner in which they operate. The basic components of windows, however, are similar in all types of window construction. A brief overview of the primary components will provide a better understanding when reference is made later to forcible entry methods on the various kinds of windows.

Glazing

The portion of the window that allows penetration of light and visibility in both directions is called the glazing. Glazing is usually made of glass, although in special applications it may be of a thermoplastic such as acrylic or butyrate. Important characteristics of glazing for forcible entry consideration are resistance to breakage, pattern of breakage, and expense of replacement. Each characteristic is dependent on the kind of glazing within the frame.

Windows designed to insulate against extremes of temperature feature factory-sealed double or triple glazing. They contain any of the below-listed types of glazing. Windows of this design have two or three sheets of glazing mounted in a frame, each separated from the next with a sealed space that contains dehydrated air (Figure 7.2). The combination of multiple layers of glazing and dead space within reduces heat entrance or loss during extreme weather conditions. The pattern of breakage will depend upon the type of glazing.

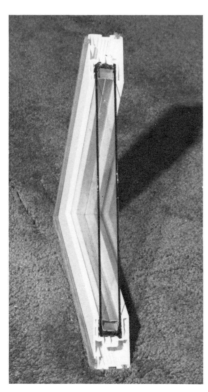

Figure 7.2 Double-glazed windows consist of two layers of glazing separated by an air space.

The most common types of glazings, along with their characteristics, are listed below:

- Clear window glass is the least expensive type of glass in use. It is available in two thicknesses: "single strength" — about 3/32-inch (2.4 mm); and "double strength" — about ⅛-inch (3.2 mm). It is commonly used in windows where high strength is not required and distortion is not a problem. Clear window glass forms knife-like shards when broken (Figure 7.3).

- Plate glass and float glass are two glasses made by different processes but that have similar characteristics. The primary difference between the two is that float glass has less distortion than plate glass. Both are commonly used in showcase or picture windows. Thickness varies from ⅛- to 7/8-inch (3.2 mm to 22.2 mm). Like the less expensive clear window glass, both break into knife-like shards.

- Tempered glass has been heat treated to increase its strength four to five times that of the same thickness of plate or float glass. Due to the extra processing required, tempered glass is more expensive than plate or float glass. Tempered glass cannot be cut or machined after tempering. It is usually identified by a small symbol etched into the glass in a lower corner (Figure 7.4). Tempered glass shatters into small, generally cubical pieces when broken (Figure 7.5).

Figure 7.5 Tempered glass breaks into small, cubical pieces.

Figure 7.3 Clear window glass forms knife-like shards when it is broken.

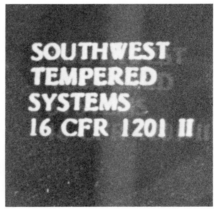

Figure 7.4 Tempered glass is usually identified by a symbol etched into one corner.

Figure 7.6 Laminated glass tends to adhere to its inner plastic sheet when broken.

- Laminated glass contains one or more layers of plastic sandwiched between two or more layers of glass. These layers add strength and impact resistance to the sheets. The glass tends to adhere to the plastic when broken, thus minimizing flying glass (Figure 7.6). Since laminated glass is more expensive than clear window or plate glass, it is usually found only where broken glass may present a safety hazard, such as in doors or next to entryways.

- Wired glass has a fine meshed wire grid embedded in it (Figure 7.7). It may be found where fire resistance or extra security is required. When broken, the pieces tend to adhere to the wire mesh.

- Bullet-resisting glass is a security glass made by bonding layers of glass under heat and pressure into ¾- to 3-inch (19 mm to 76 mm) laminated sheets. Although not totally impenetrable, bullet-resisting glass is very resistant to impact and breakage (Figure 7.8). The expense and difficulty of installation restrict the use of bullet-resisting glass to specialized applications.

- Thermoplastic glazing comes in many different types and thicknesses. Thermoplastic glazing is more resistant to impact than glass of the same thickness, so it is commonly used in areas where breakage and vandalism are problems. Three of the most common glazings are made of acrylic plastic, polycarbonate plastic, and cellulose acetate butyrate. More familiar terms for these compounds are Plexiglass© (acrylic), Lexan© (polycarbonate), and Uvex© (butyrate). The makers of Lexan claim that it is 250 times stronger than glass, with half the weight, and 30 times stronger than Plexiglass. Thermoplastic sheets are made in thicknesses up to 4 inches (102 mm), with 1/8- to 1/2-inch (3 mm to 13 mm) the most common.

Figure 7.7 Wired glass has a fine mesh wire embedded within the glass.

Figure 7.8 Although it is not totally impenetrable, bullet-resistant glass effectively resists impact and breakage. *Courtesy of Chicago Bulletproof Equipment Co.*

In terms of relative impact resistance, for the same thickness, polycarbonate is more resistant than butyrate or acrylic; butyrate is more resistant than acrylic. Thermoplastic glazing, however, will lose some impact resistance with time, particularly if subjected to direct sunlight. Ultraviolet light and heat each have a significant negative effect on tensile strength. This should be considered when thermoplastic windows are encountered during pre-incident planning. Age and condition of the glazing should be noted on the pre-incident survey form.

The framework that surrounds the glazing makes it possible to open the window, thereby allowing ventilation of a room or building. The framework also serves to maintain a weathertight seal around the window opening. This framework, with or without the glazing, is referred to as the sash. Depending on design and function, a window can have more than one sash. This is the case with the familiar double hung window, which has an upper sash and a lower sash. Each may be opened to allow ventilation through either the top or bottom of the opening. The sash is made up of horizontal and vertical frame members (Figure 7.9). The horizontal members are known as rails; the vertical members are the stiles.

Stiles and rails are made of wood, aluminum, steel, or a metal alloy. The material used is often chosen according to whether the window will be used in a commercial building or in a more residential setting. Lightweight wood and aluminum sashes are frequently used in residential windows. Steel, heavy aluminum, and hardwood are used in commercial window sashes to withstand more demanding use as well as to resist forcible entry.

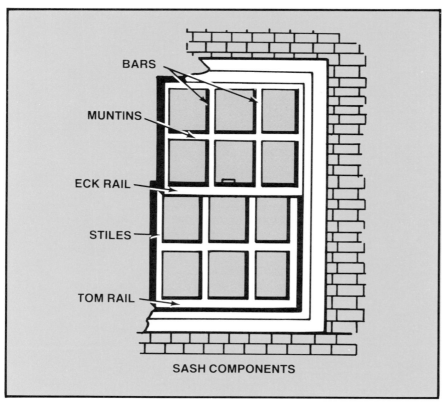

BARS

MUNTINS

ECK RAIL

STILES

TOM RAIL

SASH COMPONENTS

Figure 7.9 A window sash is made of vertical stiles and horizontal rails.

TYPES OF WINDOWS

Window designs vary widely from region to region, primarily because of established architectural styles and predominant weather conditions. For the purposes of this manual, window types are classified according to the manner in which they operate. The basic types of windows are

- Sliding
- Swinging
- Pivoting
- Security

Fixed Windows

Windows that are set in a stationary position and do not open are called fixed windows. The glazing in a fixed window may be installed in a sash mounted in the window frame. It may also be installed directly into the window frame between strips of molding. Figure 7.10 illustrates a showcase window, which most often contains plate or float glass.

Figure 7.10 Plate or float glass glazing in a fixed window may be installed in a sash mounted within a window frame, or mounted directly into the window frame between strips of molding.

Sliding Windows

Sliding windows have one or more sashes. At least one sash opens by sliding either horizontally or vertically within the window frame. Examples of sliding windows are double hung windows and horizontal sliders (Figure 7.11).

All sliding windows have tracks or channels that guide the sash when it is opened or closed. Wood-framed windows may have metal or plastic channels, or channels formed by stops or moldings attached to the frames. Metal-framed windows usually have channels or ridges molded as an inherent part of the frame.

The latching or locking mechanism for sliding windows is usually found on the rail or stile where the sashes meet (Figure 7.12). There may also be latches or bolts where the movable sash meets the frame. In addition, a common homemade device to improve window security is some sort of blocking mechanism: a wooden or metal rod placed in the opening that prevents the sash from sliding open.

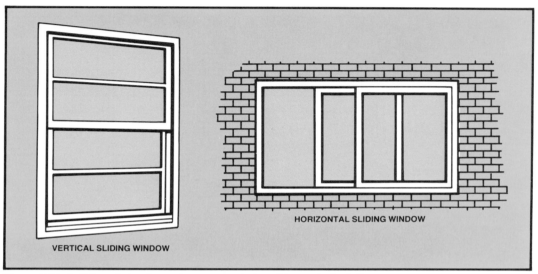

Figure 7.11 A sliding window has at least one sash that slides vertically (left) or horizontally (right).

Figure 7.12 Sliding window locks are often mounted on adjacent stiles.

Swinging Windows

Swinging windows have one or more sashes hinged on or near two adjacent corners. They may open either toward the interior (in-swinging) or toward the exterior (out-swinging). Figure 7.13

shows the glazing in swinging windows may be fully sashed, as with an awning window, or be only partly sashed, as in a jalousie window. Other examples of swinging windows include casement, hopper, and projected windows.

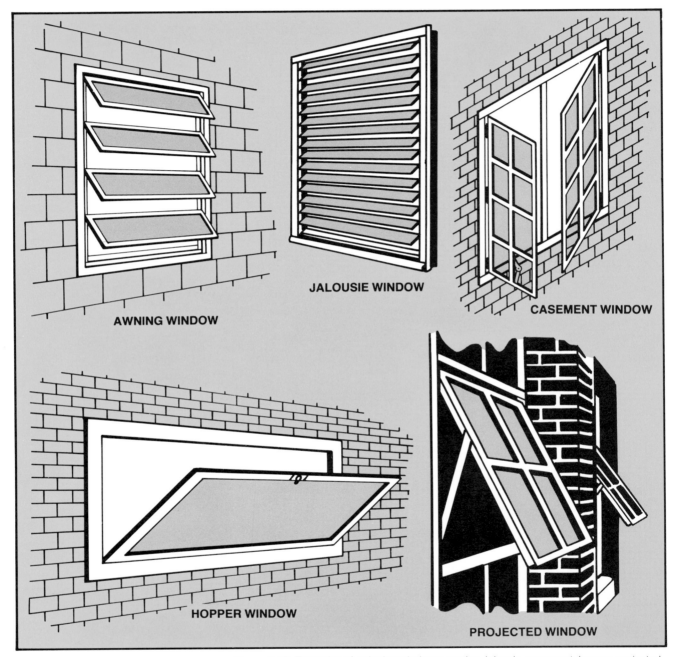

AWNING WINDOW

JALOUSIE WINDOW

CASEMENT WINDOW

HOPPER WINDOW

PROJECTED WINDOW

Figure 7.13 Examples of swinging windows: awning, jalousie, casement, hopper, projected.

Small, residential swinging windows are commonly opened manually by pushing or pulling. Larger residential and commercial window assemblies typically include mechanical opening devices such as gear-driven rotating handles (Figure 7.14) or even motorized power units. Each opening device also provides some level of security because the mechanism will resist when pressure is applied externally, as when the window is pried. Swinging windows are usually latched or locked at the portion of the sash farthest from the hinges (Figure 7.15). Surface bolts may also be added to provide additional security.

Figure 7.14 Gear-driven rotating handle for opening a commercial swinging window.

Figure 7.15 Swinging windows are usually locked at the portion of the sash farthest from the hinges.

Pivoting Windows

Pivoting windows have a pivoted or hinged sash. Horizontal pivot windows rotate on pivots attached to the center of the stiles, while vertical windows rotate on pivots attached to the center of the rails (Figure 7.16). With both types of windows, part of the sash opens toward the interior and part toward the exterior.

Pivoting windows are designed for ease of maintenance and may be cleaned from the inside. While this style of window allows excellent ventilation, it does not allow easy entrance or egress.

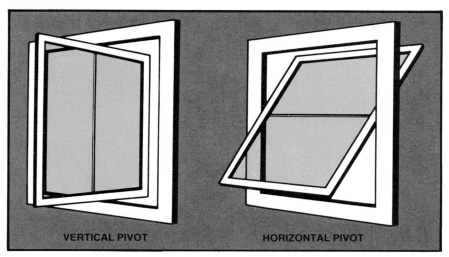

VERTICAL PIVOT **HORIZONTAL PIVOT**

Figure 7.16 A vertical pivoting window rotates on rail-mounted pivots; a horizontal pivoting window rotates on stile-mounted pivots.

Figure 7.17 Some standard windows are fitted with heavy mesh screening for additional security.

Pivoting windows have locking devices similar to swinging windows. It is becoming more common to find add-on locking devices that improve security.

Security and Detention Windows

Windows that are specifically designed to prevent *entry* are known as security windows. Windows designed to prevent *exit* are called detention windows. Both types of windows have similar features and can be treated in the same way during forcible entry.

It is common to find standard windows that have been altered to serve as security or detention windows. One way to make standard windows serve these functions is to fasten metal bars or screens to the exterior of the window frame or to the building itself (Figure 7.17). The metal bars or screens may be bolted to the building, embedded in masonry, or mounted on hinges and locked with padlocks or other locking devices. Metal bars are usually no less than 1/4-inch (6 mm) in diameter and are usually welded into complete assemblies before being mounted in or across the window (Figure 7.18).

Figure 7.18 Several types of barred windows. *Courtesy of Edward Prendergast.*

Custom-made security and detention windows may have closely spaced steel bars, straps, or mesh built into the sash. Other types use bullet-resisting glass or impact-resistant plastic glazing.

Security and detention windows may have movable sashes and some means of opening metal bars or screens. This feature allows occupants of the building to use the window as an emergency exit. Security windows may be only latched on the inside (not locked). Keys or removable handles are often required to open detention windows.

Review

Answers on page 264

Answer each of the following questions in a few words or short phrases:

1. What are three important characteristics of glazing for forcible entry consideration?

   A. _____

   B. _____

   C. _____

2. What two types of glass are commonly used in picture windows?

   A. _____

   B. _____

Complete the following statements with words or phrases that will make the statements correct:

3. The construction details of windows vary depending upon two factors:

   A. _____

   B. _____

4. Sashes in residential windows are generally constructed of either

   A. _____ or

   B. _____

5. The five basic types of windows, classified by the manner in which they operate, are as follows:

   A. _____

   B. _____

   C. _____

   D. _____

   E. _____

Determine whether the following statements are true or false. If false, state why:

6. If entry to a building has been accomplished through a door, it will not be necessary to also gain entrance through a window.

   ☐ T   ☐ F   _____

   _____

7. Jalousie windows and casement windows are examples of pivoting windows.

   ☐ T   ☐ F   _____

   _____

8. Security windows are specifically designed to prevent entry while detention windows are designed to prevent exit.

☐ T   ☐ F   _____

_____

Select the choice that best completes the sentence or answers the question:

9. A small symbol etched into the lower corner of a glass door usually identifies
A. plate glass
B. laminated glass
C. tempered glass
D. bullet-resisting glass

10. Which type of glazing contains one or more layers of plastic sandwiched between two or more layers of glass?
A. Laminated glass
B. Wired glass
C. Bullet-resisting glass
D. Thermoplastic glazing

11. A disadvantage of thermoplastic glazing is that its tensile strength is negatively effected by
A. low temperature
B. direct sunlight
C. low humidity
D. all of the above

12. Lexan and Uvex are brand names for types of
A. laminated glass
B. tempered glass
C. bullet-resisting glass
D. thermoplastic glazing

13. Double or triple glazed windows are constructed from
A. plate glass
B. tempered glass
C. laminated glass
D. any of the above

Match the type of glazing in the left column with its breaking characteristic in the right column:

Glazing	Breaking Characteristics
____ **14.** Plate glass	A. Shatters into small, generally cubical pieces
____ **15.** Float glass	
____ **16.** Tempered glass	B. Resistant to breaking
____ **17.** Laminated glass	C. Forms long, knifelike shards
____ **18.** Clear window glass	D. Adhered to plastic sheeting
____ **19.** Thermoplastic	

Forcing Windows

8

NFPA STANDARD 1001
FORCIBLE ENTRY
Fire Fighter I

3-2 Forcible Entry

3-2.1 The fire fighter shall identify and demonstrate the use of each type of manual forcible entry tool.

Fire Fighter II

4-2 Forcible Entry

4-2.1 The fire fighter shall identify materials and construction features of doors, windows, roofs, floors, and vertical barriers and shall define the dangers associated with each in an emergency situation.

4-2.2 The fire fighter shall identify the method and technique of forcible entry through any door, window, ceiling, roof, floor, or vertical barrier.

Chapter 8
Forcing Windows

Although doors are generally considered as the initial point of entry into a building, it may be necessary to force entrance through a window. In some cases, it is possible to force entrance through a window quickly and with little damage in order to open a nearby door from the inside. In other situations, the window may be the only way to reach a particular location inside a structure.

Once a decision is made to use a window for entrance, the first step is to examine the window to see if it can be opened and how it operates. Next, try to open the window in a normal fashion, without force (Figure 8.1). This step is extremely important because windows are sometimes left unlatched, making it possible to save valuable time and to avoid damage to the structure. Sometimes an unlocked window will not yield to normal opening force because it is swollen by moisture or painted into place. If the window seems "stuck," apply extra force to try to open it.

Figure 8.1 As with a door, try to open a window in the normal fashion before attempting forcible entry.

Procedures for forcing entrance through a window are varied because there are so many types and designs of windows. The most appropriate technique is best decided after a close examination of each window. In an emergency situation, there is little time available to properly assess construction features. The best time to make this examination is during a pre-fire planning inspection. This pre-incident assessment should include identification of the best points of entrance, possible methods to gain entry, and tools required to carry out the operation.

SIZING UP THE WINDOW

Whether size-up takes place during the pre-fire planning inspection or during an emergency situation, it is a vital part of forcible entry activities. A primary concern during forcible entry is the safety of personnel. Firefighters can minimize the dangers of getting cut by broken glass or being otherwise injured by following accepted methods and practices during forcible entry operations.

It is also necessary to consider the effect of opening a window in a building that has fire inside. If signs indicate that there is potential for a backdraft, notify the incident commander before proceeding so that personnel can initiate preventive measures such as vertical ventilation. If opening a window is likely to increase the intensity of the fire in the room, consider the need for hoselines or exposure protection.

Because some methods of forcing entry are faster than others, the amount of time available may be a primary factor in choosing the entry technique. Quick entry for a rescue operation may require breaking a window, while a less urgent situation may allow personnel to disassemble the window, resulting in less damage. A number of factors influence the time needed to force entry. For instance, breaking and clearing wire-glass from a sash takes more time than breaking and clearing out ordinary window glass. Some latches are more secure than others, and may take a longer time to force. Unlatching a window only to find that it is opened by a mechanical assembly and cannot be forced by hand can waste valuable time. It is important to identify all such construction features and understand how they affect the time required for entrance.

Forcing a window must provide enough access to perform necessary work. In most situations, the window openings must provide enough clearance for a firefighter wearing full protective equipment, including breathing apparatus, to enter (Figure 8.2). If a number of personnel with equipment must use the opening, the size-up process should also take this into account. Using a window as the primary entry point for a large number of people

Figure 8.2 Choose window openings that provide enough room for a fully equipped firefighter to crawl through.

may cause unacceptable delays in operations if the opening is too small or difficult to reach.

Another factor to consider during size-up is the replacement cost of windows that are partially or totally destroyed by forcible entry. It is preferable to minimize damage, thereby keeping repair costs low. Knowing the relative cost of different types of glass or the amount of damage that may be caused by forcing a window sash helps in the selection of appropriate techniques.

There are other considerations to make when considering window entry that are not necessarily associated with economics. Windows constructed with stained glass or other special features may have historic, artistic, or sentimental value. These windows are found in churches, assembly halls, and historical buildings. When a window of this type is destroyed, the community loses more than money. Do not force entry through such windows by breaking glass *unless absolutely necessary*. Identify alternative routes and means of forcing entry during pre-incident planning.

The major factors to consider when sizing up windows are as follows:

- Requirements of situation
- Time factor
- Safety of personnel
- Type of glazing
- Type of latch or lock
- Type of opening mechanism
- Costs of repair or replacement

The following descriptions of forcible entry techniques are intended to help personnel select an appropriate technique for each window type. An understanding of window construction and of the techniques presented will help in the decision process.

Regardless of the technique used, always consider *safety* first when performing forcible entry. ALWAYS wear personal protection equipment when forcing entry through windows. This includes face and eye protection, gloves, and full body protection. While turnouts and gloves help protect the body, they are not absolutely impenetrable to glass. Such protection helps guard against cuts from small pieces of broken glass that fly from the window if and when the glass breaks. These small projectiles may be carried long distances by the wind.

FORCING FIXED WINDOWS

Since fixed windows have no movable parts, there are only two ways to gain entry: either remove (disassemble) the glazing from the frame or break the glazing out completely. Disassembling the window usually results in a lower replacement cost but takes longer than breaking out the glazing. Deciding which entry method to use will be affected by an evaluation of the time available versus replacement costs.

Disassembly of a fixed window is usually restricted to wood-framed windows. If the glazing is held in the frame with molding, it may be possible to use a chisel, screwdriver, or other suitable tool to pry the molding away from the glazing. Prying is also possible when glazier points and caulking have been used to set glass in a sash. Once molding or caulking has been removed, the glazing can be lifted out in one piece. A large piece of glazing can be awkward to handle and usually has sharp edges, so take care to avoid dropping the glazing or being cut when handling large sheets. **CAUTION:** Always wear gloves and eye protection when handling glazing.

Breaking the glazing may be the best method if rapid entry is required. It is important to determine the type of window glazing in order to know what to expect when trying to break it.

If the glazing is glass, knowledge of the type of glass provides a good idea of the pattern of breakage. For instance, tempered glass shatters into small cubical pieces, while plate glass and float glass generally form long, knife-like shards when broken. Whenever forcible entry calls for breaking glass, follow safe procedures to avoid possible injury. Piles of broken glass make unstable and unsafe footing. Avoid stepping in or walking through broken glass, if possible. If it is necessary to walk over glass, use extreme caution.

Breaking Glass

The steps for safely breaking glass out of doors or windows, regardless of the type of glass, are as follows:

Step 1: Select the appropriate tool, such as an axe, to break the glass. **CAUTION:** Do not use such equipment as a helmet or handlight for this task. UNDER NO CIRCUMSTANCES SHOULD THE HANDS BE USED, NO MATTER HOW WELL PROTECTED.

Step 2: Take a position to the side of the window, on the upwind side. If there is fire in the interior, staying upwind keeps the firefighter safely out of heat and smoke that will be released through the opening when the glass is broken.

Step 3: Incline the tool so that the hands are above the portion of the tool used to break the glass. This will prevent glass shards from sliding down the tool toward the body.

Step 4: Strike the glass sharply. If an axe is used, strike the glass with the flat side of the axe head (Figure 8.3). The point of impact should be as high on the window as possible. This minimizes the amount of broken glass that could fall toward the firefighter.

Step 5: Starting at the top of the opening, use the tool to clean out the window sash, removing any pieces of glass that remain. Cleaning out the sash makes entrance through the opening much more safe.

Figure 8.3 With the handle inclined above the axe head, strike the glass with the flat side of the axe.

Breaking Thermoplastic Glazing Loose

If the glazing is a thermoplastic, breaking is not the best method unless it is 1/8-inch (3 mm) or less thick. It is usually better to remove the glazing from its frame by disassembling it. If rapid entry is required, batter the glazing from the frame with sledges or cut it with saws. Thermoplastics such as "Lexan" and "Plexiglas" are found in both fixed and openable windows. Built to withstand extreme impact, these units are best handled in other ways than breaking.

In some installations, such as schools, thermoplastic-glazed windows are actually assembled to give way to fire fighting tools such as axes or sledges. Molding is set so that a few well-placed blows will dislodge the glazing from its sash. When using this method, it is important to select a battering tool that is heavy enough to break the glazing loose with a minimum number of blows. A sledge is ideal for this task: use one weighing no less than 12 pounds (5.4 kg).

Swing the sledge in a controlled manner so that blows can be made accurately. Avoid overswinging. Let the weight of the sledge head, rather than high velocity, accomplish the task.

The first point of impact should be at a corner of the glazing. After making the first blow, move progressively around the sash, placing succeeding blows along the edge to continue breaking the remaining molding loose.

Cutting Thermoplastic Glazing

One of the most effective means of penetrating a window with Lexan©, Uvex©, or Plexiglas© glazing is to cut it with a circular saw. The size of cut will depend on window type and the intended use of the opening. If the window is of the fixed type (no moving parts), cut an opening large enough to allow passage of fully equipped firefighters. If the window can be opened, it may be only necessary to cut a hole large enough to allow reaching in and unlocking the internal lock.

Using a circular saw with a carbide-tipped blade will give best results when cutting thermoplastic glazing. Select a blade with a medium number of teeth. When making the cuts, apply the blade firmly to the glazing, but do not rush progress of the blade. Expect that cutting speed will decrease with thicker plastics. With heavier glazing, there is a tendency for the plastic to melt from the friction of the blade and fill in the space behind the blade. If the glazing begins to melt, apply water lightly as the cut is made to cool the blade and plastic.

WARNING
Because of the electrical hazard, however, do not use water behind an electric saw.

FORCING SLIDING WINDOWS

The first step to take in forcible entry of any movable window is to try to open it in a normal fashion. This action avoids unnecessary damage caused by forcing the window and reduces the time required to gain entry.

Windows that open by sliding, such as double hung windows and horizontal sliders with either wooden or aluminum frames, are commonly found in both commercial and residential occupancies. Techniques for forcing these windows include loiding the lock, prying the sash, breaking the glazing to unlock the window, and breaking glazing out of the entire sash.

Loiding is a technique that involves the use of a thin, flat object to manipulate and open the lock. Loiding causes little or no damage to the window, but success depends on sash design, location and type of lock, and the presence of additional locking devices. There is no standard method for loiding because there are so many window and lock designs, but in all cases the loiding tool is inserted between the sashes near the lock and manipulated into position to either push or pull the device into the unlocked position. Once the lock is thrown, slide the sash open.

Prying a sliding sash usually requires breaking the locking mechanism because of its interlocking design. This type of lock has a component mounted on each sash that joins when the lock is pushed or turned into the closed position. With wooden frames, prying on the frame may cause the screws holding the lock to pull out of the wood, freeing the window to open. With aluminum or steel frames, it is unlikely that the lock will separate from the sash. When extreme prying force is applied, a rail or stile is more likely to warp or buckle, or the glazing may break.

To pry the sash open, use an appropriate pry tool. This means a tool that is tapered enough to slide between sashes or between the frame and sash, as well as one that transfers sufficient force to the prying end when pressure is exerted to the handle. Place the adz end of the tool between the window frame and the center of the sash at a point opposite the lock (Figure 8.4 on next page). Pry the sash with steady force, expecting that the lock or glazing may break suddenly. If glazing breaks, stop prying and reach through to unlock the window, then slide it open.

Since standard sliding windows are relatively easy to open even when locked, it is common for owners to use additional locking mechanisms such as surface bolts, push pins, or simple bars placed between sashes and frames. If it is impossible to open the sash because of any of these devices, it may be necessary to break the glazing from the window.

Before breaking the glazing, determine what size opening is needed to permit reaching in and unlocking the window. In some cases, it may be possible to unlock and open the sash by breaking only a small pane of glass. If it is necessary to break out the entire glazing in the sash, the sash should still be unlocked and opened before entrance is made to decrease the hazard to personnel (Figure 8.5). It will also prevent further damage to the window unit. Follow the technique for breaking glass described in the section on fixed windows.

Figure 8.4 To pry a sliding window, insert the prying tool adz between the frame and sash near the lock and pull back to break the lock.

Figure 8.5 After removing the glazing from a sliding window, unlock and open the sash to prevent damaging the sash when climbing through the opening.

FORCING SWINGING AND PIVOTING WINDOWS

As in sliding windows, swinging and pivoted windows may be forced by breaking the glazing, loiding, or prying to force the lock. Most of these methods require that personnel know where the latches are located on the window sash. On swinging windows, latches are usually found on the side opposite the lock. On pivoted windows, latches are usually located on a side that does not have a hinge pin.

As with sliding windows, it may be possible to break a small pane of glazing, unlatch, and open the sash to allow entry. In many cases, this technique allows quick entry and results in less damage than other forcible entry techniques. In some cases, however, after unlocking the window, it may still be necessary to operate a mechanical opening device to open the window. Opening mechanisms for swinging windows vary greatly. Some open manually by pushing the window out with a bar or rod. Other windows use a manually operated crank mechanism. In industrial applications where multiple window units are mounted in groups, they may be opened with manually or electrically operated opening mechanisms. It may be necessary to operate the mechanism to open these windows. If it is not possible to operate the window opener because of inaccessibility or loss of electrical power, breaking the entire glazing out is probably the best alternative. The same steps used for breaking the glass out of sliding windows apply when breaking glass out of swinging and pivoted windows.

Although loiding may be successful in some cases, many of the well-constructed commercial window units are built in a way that prevents loiding. Prying is another alternative that is easiest on wood-framed windows. Metal-framed windows usually bend, spring, or are otherwise damaged by prying.

Small awning windows and jalousie windows present different problems. It may not be possible to enter through small awning window frames, even after the glazing is broken out or the window is opened. Jalousie windows may have glazing that is easy to break, but the cost of replacing the glass can be prohibitively expensive. It may be necessary to search for other avenues of entry in these situations.

FORCING SECURITY AND DETENTION WINDOWS

Standard window units are often the weakest point in the security of a building. For this reason, owners often install special security windows or modify standard windows to provide greater security. Firefighters may also encounter window units that are designed to prevent exiting through the window, as are found in detention facilities. In some cases, entry through these windows may prove too time consuming to be practical. In other cases, application of intelligently applied force may allow entry in a reasonable time.

As with standard windows, many types of security windows are available. For this reason, accurate size-up that identifies the window type is essential to determining the best forcing method. Pay special attention to the way each window is mounted in the frame and, if window bars are used, to how the bars are anchored. Bars and security screens mounted over windows may be permanently installed, hinged at the top or side, fitted into brackets, or securely locked to the window frame or wall of the building. The type of attachment will dictate the methods to be used for entry.

To free bars embedded in masonry, use one of the following methods:

- Strike the bar with a heavy sledge directly at the point where the bar meets the masonry. Fracture lines in the

Figure 8.6 Strike the bar with a heavy sledge directly at the point where the bar meets the masonry (if available, also use a hammer-head pick, as shown).

masonry will appear after several blows and eventually the masonry should crumble away, freeing the bar end (Figure 8.6).

● Attach a tow chain or cable to the bar assembly. With a vehicle or a winch, pull outward with steady pressure. This method may be improved by weakening anchor points with a sledge or hammer-head pick.

● Cut the bars near the anchor points with a cutting torch. On windows with barriers anchored on opposite sides, make cuts on one side only, then bend the assembly to the anchored side. On windows with anchor points on all four sides, make cuts on three sides and bend the assembly to the anchored side. Avoid heat contact with glazing, wooden sashes, and frames.

● Pry bar or screen assemblies with a hydraulic power tool. Position tips directly adjacent to an anchor point with the fulcrum tip against a solid point on the wall.

● Pry bar or screen assemblies with an air bag. Position the bag between anchor points, making sure that it is placed so that when inflated it will be supported by the wall (Figure 8.7). Avoid contact with the window and glazing.

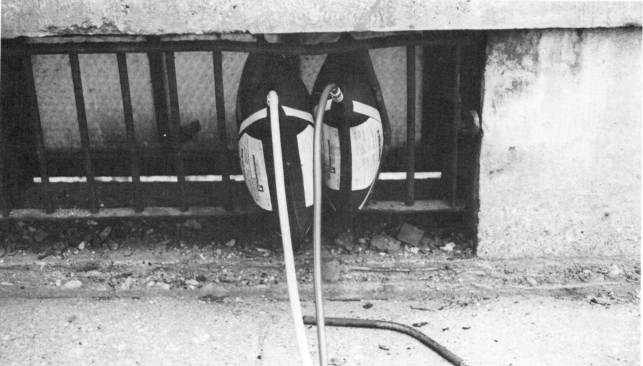

Figure 8.7 Air bags can be used to break bars loose from masonry. *Courtesy of Paratech Inc.*

SPECIAL PROBLEMS

Energy conservation concerns have given rise to many window modifications designed to cut energy losses. Double glazing is commonly encountered, sometimes in the form of storm windows, which are considered a fixed type of window. Storm windows consist of a simple sash and glazing mounted on the exterior or interior of existing windows. Exterior storm windows can usually be removed easily with screwdrivers or prying tools because they are not designed to prevent entry.

Insulating shutters and blinds may also present an additional layer of protection that must be penetrated during forcible entry. Again, these devices are not designed to prevent entry and can thus be pried open with little effort.

This discussion of entry methods assumes that the work takes place at ground level. If it is necessary to use forcible entry techniques at higher elevations, entry procedures require consideration of such details as ladder placement and falling glass. Ladders should be placed so that personnel working at a window are not exposed to smoke and flame when a window is opened or glazing is broken. Usually, ladders should be placed to the upwind side of the window, with the top of the ladder contacting the building at a point high enough to allow work to be done properly. Always anchor to the ladder with the leg lock method, with a ladder belt hook, or with an approved ladder-anchoring device.

When breaking glazing from a ladder, stand high enough above the impact point to allow the hands to be above the tool head. As when working at ground level, this technique prevents glass shards from sliding down the tool handle toward the body.

The higher the window in glass-breaking operations, the more likely that glass fragments will fly further from the building during descent to the ground. Clear the area below to prevent injury to personnel. Anchor the ladder to the building with ropes or ladder straps so that personnel will not be required to stand at the ladder base and thus be exposed to falling glass.

Answer each of the following questions in a few words or short phrases:

Answers on page 264

1. What three factors regarding forcible entry to the building should be included on a buildng pre-fire plan?

   A. _____

   B. _____

   C. _____

2. What criteria should be used to determine whether or not a window should be forced?

   A. _____

   B. _____

   C. _____

   D. _____

3. What are the two methods of entry for fixed windows with glass glazing?

   A. _____

   B. _____

4. When using an axe to break the glass out of a window,
   A. . . . on which side should you stand?

   _____

   B. . . . which part of the axe should strike the glass?

   _____

   C. . . . where on the window should the point of impact be?

   _____

   D. . . . where should the hands be held in relation to the axe head?

   _____

5. It has been decided to enter a building through a sliding window. What should you do *first* to make entry?

   _____

   _____

6. What are the four basic techniques for forcing sliding, swinging, and pivoting windows?

   A. _____

   B. _____

   C. _____

   D. _____

7. The latch is usually on which side of a

   A. swinging window? _____

   B. pivoted window? _____

Determine whether the following statements are true or false. If false, state why:

8. Walking on broken glass should be avoided even if full turnout gear is worn.

   ☐ T ☐ F _____

   _____

9. An effective means of penetrating a window with "Lexan" glazing is by cutting with a cutting torch.

   ☐ T ☐ F _____

   _____

10. Glass can be broken out of a window with the hands if gloves are worn and the face and eyes are well protected.

    ☐ T ☐ F _____

    _____

11. If the entire glazing has been broken out of a sash, the sash should still be unlocked and opened before anyone enters through the window.

    ☐ T ☐ F _____

    _____

12. The recommended method of forcible entry through storm windows is by breaking the glazing to access the lock.

    ☐ T ☐ F _____

    _____

Select the choice that best completes the sentence or answers the question:

13. Which of the following are generally considered the initial point of entry to a building?

    A. Doors   B. Windows   C. Walls   D. Roofs

14. "The access provided by forcing a window must be adequate for the job to be done." This statement means that the

    A. window should be as close as possible to the fire

    B. window cannot be above ground level

    C. window opening should provide room to allow passage of a firefighter wearing breathing apparatus

    D. window should not be forced open if there is an alternative means of entrance to the building

15. Which is the most important consideration when assessing alternative forcible entry techniques for a window?
    A. Time factor
    B. Safety of personnel
    C. Access requirements of the situation
    D. Type of window components involved

16. Which would usually be the best way to effect forcible entry through a thermoplastic window in a school?
    A. Loiding
    B. Breaking out the glazing
    C. Prying to force the lock
    D. Cutting the glazing with a circular saw

17. Prying would normally be most effective on a sliding window with
    A. an aluminum frame
    B. a steel frame
    C. a wood frame
    D. none of the above

18. When bars are mounted over a window, the method used for entry through the window will depend on
    A. the size of the window
    B. the spacing of the bars
    C. how the bars are anchored
    D. how the window is mounted in the frame

19. Window bars imbedded in masonry can be freed by
    A. striking the bar anchor points with a heavy sledge
    B. pulling the assembly free with a winch and cable
    C. prying the bars with an air bag
    D. any of the above

20. When using a ladder to reach a window above ground level so that glazing can be broken out,
    A. the top of the ladder should be directly below the window
    B. the bottom of the ladder should be anchored by a firefighter
    C. the firefighter on the ladder should attach a ladder anchoring device
    D. all of the above

Through-
the-Wall
Entry

9

**NFPA STANDARD 1001
FORCIBLE ENTRY
Fire Fighter I**

3-2 Forcible Entry

3-2.1   The fire fighter shall identify and demonstrate the use of each type of manual forcible entry tool.

Fire Fighter II

4-2 Forcible Entry

4-2.1   The fire fighter shall identify materials and construction features of doors, windows, roofs, floors, and vertical barriers and shall define the dangers associated with each in an emergency situation.

4-2.2   The fire fighter shall identify the method and technique of forcible entry through any door, window, ceiling, roof, floor, or vertical barrier.

Chapter 9
Through-the-Wall Entry

The first eight chapters of this manual focus on the familiar points of entry in most structures. These components — doors, windows, and locks — are usually targeted during forcible entry because they are more vulnerable to entry by force than any other components of the structure. Doors are the logical first choice because, by design, they make passage in and out of the building an easy matter, once they are opened. Through-the-lock entry is also an effective means of forcing doors with very little damage to even the lock itself. Window entry is popular because glass can be quickly broken to allow access, usually so doors can be unlocked from the inside.

There are situations, however, in which conventional entry methods are not appropriate. This would be the case, for example, in a windowless building with few doors, such as a concrete-walled warehouse. A fire in the front of this type of structure might require that entry be made through a rear wall to stop the fire from extending to unaffected areas.

A primary concern when considering the feasibility of through-the-wall entry is whether structural integrity will be affected. To properly size up the situation, there should be an understanding of the relationship of walls to the rest of the structure. Walls serve not only to protect the inside of the structure from the elements, but are also an inherent part of building stability. Walls help support the weight of structural components such as floors, ceilings, and rafters, as well as the weight of the contents of the building. Walls may also contribute to, or resist, the spread of fire.

This chapter discusses the fundamental construction features of both exterior and interior walls, and stresses the importance of recognizing wall function when determining whether through-the-wall forcible entry is feasible. Moreover, load-bearing and fire-resistive characteristics of walls are considered to be an inherent aspect of wall function and firefighter safety, and must be taken into consideration when forcing operations are initiated.

Walls can be broadly categorized by the materials used in their construction. Masonry walls are built of concrete, concrete block, brick, stone, or combinations of these materials. Metal walls are made of steel, aluminum, or alloys. Wood-framed walls are composed primarily of dimensioned lumber of varying sizes.

WALL CONSTRUCTION

A building may be structurally supported in one of two ways. The total weight of the structure, including both live and dead loads, can be supported either by the walls or by a structural frame. Both methods are widely used. The feasibility of either method is usually determined by the height of the structure. Buildings of two or three stories generally have walls as the main supports (Figure 9.1). Taller buildings, however, because of the increased load-bearing demands that come with increased height, require a structural frame (Figure 9.2). A notable exception to this rule are older masonry buildings that have structural supporting walls of up to 70 feet (21 m) high. While this construction practice is no longer permitted, many of these old structures are still in use and should be identified during pre-fire surveys.

Figure 9.1 The walls are usually the main structural supports of two- or three-story buildings.

Figure 9.2 Because of increased load-bearing demands, multistory buildings are built on structural frames.

When a wall is used for structural support, it is termed "load bearing." Masonry buildings, either ordinary or heavy timber, utilize load-bearing walls as the primary structural support.

Some interior walls may also be load bearing. If the interior walls serve only as partitions, interior structural loads are supported by columns (Figure 9.3).

The structural frame can be made of steel, concrete, or wood. When a frame provides structural support, exterior walls only enclose the building and are termed curtain, or panel, walls. Curtain walls are made of such materials as glass, metal, tile, concrete block, brick veneer, wood, and various sidings.

While walls are fundamental structural components of buildings, they also affect a building's fundamental fire safety. Voids within walls provide paths for the communication of fire, and combustible walls contribute significant fuel to the fire. Fire-resistive walls, on the other hand, block the spread of fire both internally and externally.

Figure 9.3 In some high-rise buildings, interior loads are supported by columns.

Masonry Walls

An exterior masonry wall is one of the simplest load- bearing walls. Constructed of concrete, concrete block, stone, brick, or combinations of these materials (Figure 9.4), all perform well when exposed to fire, having a fire resistance rating of two to four hours.

Masonry walls in multistory buildings are often wider at the bottom than at the top. A multistory masonry building may have walls 20 inches (508 mm) thick at the base and only 6 inches (152 mm) at the top.

Variations in cement, mortar, reinforcement components, and workmanship affect the fire resistiveness and load-bearing capacity of each type of wall. If a masonry wall is structurally sound, failure during a fire usually occurs only because force is exerted against the wall by collapsing interior components such as ceiling joists and rafters (Figure 9.5). For this reason, masonry walls are typically the last structural component to fail in a burning wood-joisted masonry building.

Fire-resistive exterior walls do more than provide structural stability; they also tend to reduce the communication of fire from one structure to another structure. Building codes usually require less clearance between buildings with masonry or fire-resistive exterior walls than between buildings with combustible exteriors. Large window openings, however, can still present exposure problems.

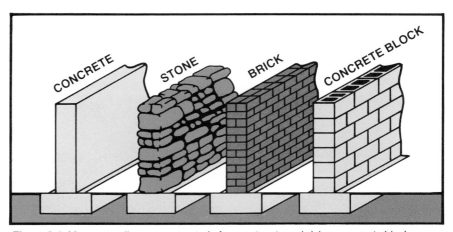

Figure 9.4 Masonry walls are constructed of concrete, stone, brick, or concrete blocks.

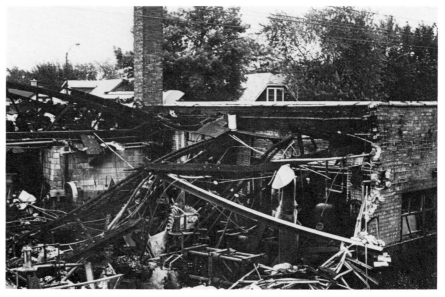

Figure 9.5 The failure of a masonry wall during a fire usually occurs because of the force of collapsing interior components falling against the wall.

Concrete Walls

Some exterior walls are formed of precast concrete. These walls are made of slabs that are 5 to 10 inches (127 mm to 245 mm) thick, often cast at the building site. They are cast in a horizontal position and, after hardening, are lifted into position with a crane

(Figure 9.6). Sometimes the slabs are tipped up on their bottom edges, from the horizontal to the vertical plane, giving rise to the term "tilt-up construction."

Precast wall construction may be either load bearing or non-load bearing, with either solid or "sandwich" panels. Sandwich panels, made with a one- or two-inch (25 mm or 51 mm) layer of polystyrene placed between the inside and outside layers of concrete, provide better thermal insulation than solid panels.

Tilt-up concrete construction is highly fire resistive. However, since a single panel can weigh 20 tons (18 metric tons), its collapse under fire conditions poses a severe threat to firefighters. Forcible entry operations through concrete walls should only be attempted when it has been established that fire has not seriously reduced the stability of wall supports.

Figure 9.6 Precast concrete walls are lifted into position with a crane.

Concrete Block Walls

Concrete block has several classifications, including load bearing and nonload bearing, hollow and solid. A concrete block that has a cross sectional area, 75 percent of which is solid material, is classified as solid block. Aggregate materials used to make concrete blocks include gravel, crushed stone, air cooled slag, cinders, and shale. The fire resistiveness of concrete blocks varies according to the thickness of the block and the aggregate used in the concrete. Concrete blocks made with pumice, a light volcanic rock, have the highest fire resistance rating.

Brick and Stone Veneered Walls

A wood-frame building may have an exterior facing of brick (Figure 9.7 on next page). Such construction is termed "brick veneer." Brick veneer construction provides the architectural styling

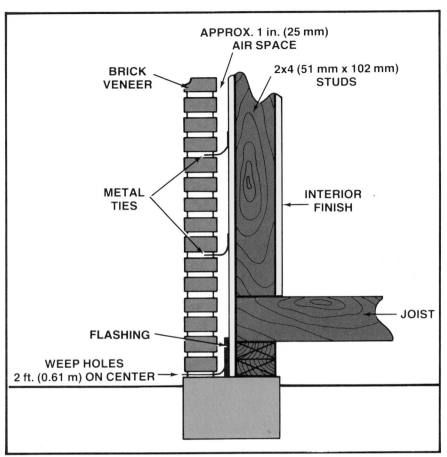

Figure 9.7 A wood-frame building may be faced with a brick exterior called a "brick veneer."

of brick without the cost. The brick veneer adds little to the structural support and must be tied to the wood-frame wall at intervals of 16 inches (406 mm). However, brick veneer does add to the thermal insulating value of the wall.

The external brick layer protects a frame structure from external exposure to fire. However, since the main structural support is still provided by an internal wood frame, there is little difference between a brick veneer building and an ordinary wood-frame building with regard to an internal fire.

From the outside it is difficult to determine visually if a building has brick bearing walls or brick veneer walls. When forcible entry is attempted through brick walls, however, it is reasonable to expect a wood-frame wall behind the brick facing. For this reason, proper tools should be selected to breach both types of materials before starting the operation.

Metal Walls

Metal is frequently used today for exterior walls, especially in commercial buildings. Prefabricated metal wall members are becoming common, replacing wooden studs in many types of structures, including hospitals, office complexes, high rises, stores, and even service stations. The metal usually comes in sheets, sections, or panels that fasten to wood or metal studs with bolts, screws, rivets, or welds. The metal may be either painted or coated with porcelain.

It should be noted that metal walls tend to be expensive to repair if they are damaged during forcible entry. Curtain walls, for example, are often composed of large, prefabricated panels. Even if only a small section of the panel is damaged, it is usually necessary to replace the entire panel.

Wood-Frame Walls

Perhaps the most common type of wall construction is that in which the structural support is provided by a framework of wooden members. Both exterior and interior walls support interior structural components, such as floor joists and roof rafters. Wood-framed walls are made of 2 x 4-inch (51 mm x 102 mm) or 2 x 6-inch (51 mm x 153 mm) vertical studs placed 16 to 24 inches (406 mm to 610 mm) on center. The space between the studs is either hollow or filled with insulation, but in either case the space may be blocked with fire stops.

Wood-frame structures are of two general types: balloon frame and platform frame. In a two- or three-story balloon-frame structure, the exterior wall studs are continuous from the foundation to the roof (Figure 9.8 on next page). The first floor joists are nailed directly to the studs. In a platform-frame structure of similar height, exterior wall studs are not continuous from the first floor to the roof. The first floor is constructed as a platform upon which the exterior wall studs for the first story are erected (Figure 9.9 on next page). A double layer of 2 x 4-inch (51 mm x 102 mm) or 2 x 6-inch (51 mm x 153 mm) members, called plates, are laid horizontally along the top of the studs. The second-story framing is then erected on top of the "platform" formed by the first-story ceiling joists and supporting walls. Some interior walls also support this platform load.

A major hazard with balloon-frame construction is that the vertical spaces between studs in exterior walls provide a channel for the rapid communication of fire from floor to floor. Fire can travel vertically through the wall and into horizontal floor joists, and quickly engulf all floors. The only check against this rapid fire spread is horizontal fire stops nailed between the studs, but this feature is not consistently found in balloon construction. For this reason, a fire in this type of structure is much more difficult to control than one in a platform-frame building, where wall plates serve as fire stops.

Exterior walls are usually covered with several layers of materials. Sheathing is installed on the outside of exterior wood-

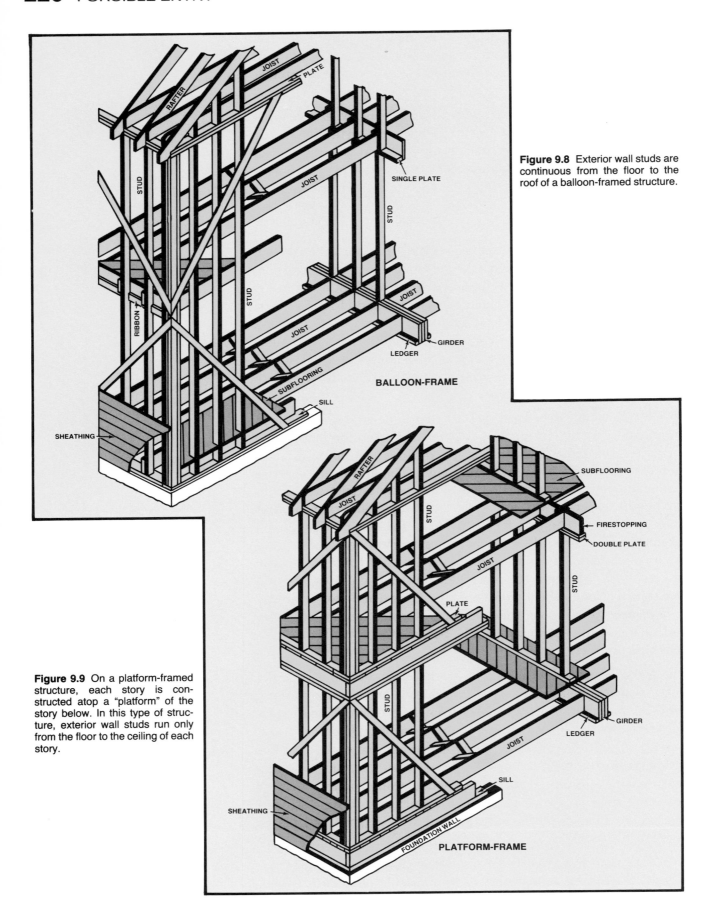

Figure 9.8 Exterior wall studs are continuous from the floor to the roof of a balloon-framed structure.

BALLOON-FRAME

Figure 9.9 On a platform-framed structure, each story is constructed atop a "platform" of the story below. In this type of structure, exterior wall studs run only from the floor to the ceiling of each story.

PLATFORM-FRAME

frame walls (Figure 9.10) to provide structural stability and weatherproofing, as well as to provide an underlayer for an exterior wall covering such as siding. One of the most common sheathings is plywood, and there are also a number of fiberboard materials used. To further weatherproof a structure, building paper is stapled either under or over the sheathing. This heavy-duty paper is treated to minimize air movement through the wall.

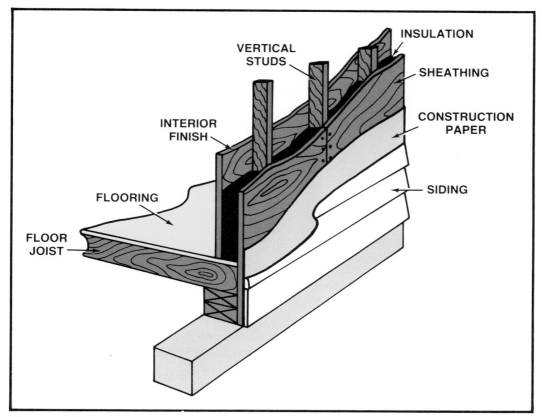

Figure 9.10 Sheathing is installed on the outside of exterior wood-frame walls to provide structural stability, weatherproofing, and to serve as an underlayer for the exterior finish material.

Siding material, available in a wide range of architectural styles, can include wood clapboards, board and batten or aluminum siding, stucco, and shingles of wood, asphalt, or asbestos cement (Figure 9.11 on next page). Brick, stone, and other masonry materials are also used as exterior wall coverings and also come in a wide range of styles.

The interior surface of the wall is almost always gypsum or plasterboard, although in older buildings it may be lath and plaster. Interior surfaces also may be covered with wooden paneling or veneers, which are usually attached to the wall covering with nails and/or adhesive.

Since the early 1970's, a developing interest in energy conservation has prompted increasing numbers of builders and homeowners to insulate building walls to reduce heat loss to the

Figure 9.11 Exterior finishes include such materials as wood clapboards, stucco, aluminum siding, and even shingles.

outside. Insulation is accomplished by filling the empty spaces between wall studs with a material that resists conduction of heat (Figure 9.12). Various materials are available for use as insulation in exterior wood walls, including mineral wool, corkboard, fiber glass, shredded paper, wood fiber, polyurethane, and polystryrene. All of these materials are available in several forms, including rigid boards, batts, blankets, and loose fill.

Although a well-insulated building is beneficial to saving energy, it has an undesirable affect on fire behavior. Because it prevents the escape of heat from the building interior, the rate of fire acceleration is much greater than in an uninsulated building of similar construction, and flashover is likely to occur sooner. This is because any insulated space retains more of the thermal energy released by a fire within it, thus the temperature within the space rises more rapidly.

The use of foam plastic as an insulating material has attracted considerable attention in recent years. Because foam plastics are combustible and because of the rapidity with which flame spreads over their surface, building codes impose stringent regulations on their use. Typically, codes require that this insulation be faced with a thermal barrier, such as gypsum wallboard, to prevent or at least retard surface ignition of the foam.

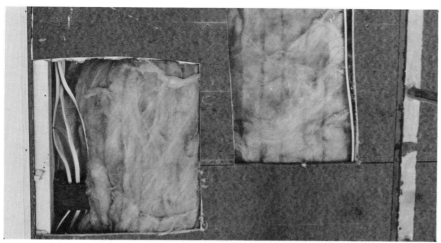

Figure 9.12 Walls are insulated by filling all voids with a material that is a poor conductor of heat.

The extent to which the presence of a foam insulation in a wood-frame wall increases fire spread within the wall depends on the presence of an air space between the foam and the wall. If an air space exists between the insulation and the wall surface, fire development within the wall is rapid, because the fire spreads over the plastic surface and has air available from within the space. If, however, the space is completely filled with foam, the fire burns upward through the material, but progressing more slowly than if the space was empty. This is especially true if the foam is sandwiched between noncombustible coverings.

The use of a combustible insulation in walls also increases the possibility of a fire starting within the wall. This is due to the possibility of a spark from electrical wiring and connectors igniting the insulation.

Loose-fill insulation may be made of such materials as granulated rock wool, granulated cork, mineral wool, and glass wool. The loose insulation can be either blown into stud spaces or packed in by hand. Cellulose fiber and shredded wood can also be used as loose insulation material. Although they can be treated with water-soluble salts to reduce their combustibility, such materials still burn in a slow smoldering manner. Whenever a fire enters a wall space, therefore, good fire fighting tactics require that the wall be opened and that the insulating material be thoroughly checked.

If the exterior wall of a multistory building is a curtain wall, a potential for vertical communication of fire exists. Strictly speaking, a curtain wall is any exterior wall that is not load bearing (or supports only its own weight) and therefore can include some masonry exterior walls such as brick and clay tile. However, the term is most commonly applied to various prefabricated glass and metal assemblies. These materials provide almost no fire resistance. In some designs, the curtain wall is entirely or almost entirely made of glass, extending from the floor to the ceiling of a given story (Figure 9.13). If a fire develops on one floor of a building with glass curtain walls, the heat cracks and breaks the glass. If there is a sufficient volume of fire within the room, the flame plume projects through the broken window and overlaps the ceiling slab above. The flame can then expose the windows of the story above, causing them to break, and allow the fire to enter the upper floor.

When a curtain wall consists of metal and glass panels, the method of attachment to the structural frame is important to firefighters. It is possible for a small open space to exist between the curtain wall and the floor slab that permits vertical extension of fire up into the inside of the curtain wall. To prevent this, suitable firestopping must be provided to maintain the continuity of the floor as a fire-resistive barrier. Curtain walls are vulnerable to forcible entry operations because they are not built to bear weight. It is still important, however, to consider that there may be fire behind the wall when opening it for entry. Always position hoselines before starting opening operations.

Figure 9.13 Some curtain walls are made primarily of glass, extending from the floor to the ceiling.

Interior Walls

Interior walls may or may not be load bearing, depending on the need of the wall to support the weight of overhead ceiling and roof spans. As with exterior walls, interior load-bearing walls are built to support weight. Interior walls are known as partition walls if they are not load bearing. Fire resistance requirements for both kinds of walls are usually determined by the local building code. For example, the partition wall separating adjacent apartments in an apartment building may be required to have a one-hour fire resistance rating. This requirement protects the occupants of one apartment from a fire in a neighboring apartment.

A common way to provide fire resistance is to use ⅝-inch (16 mm) fire-rated gypsum wallboard applied to both sides of 2½-inch (64 mm) steel studs. Such an assembly could not be used as a load-bearing wall. However, if ⅝-inch (16 mm) gypsum wallboard was applied to both sides of a 2 x 4-inch (51 mm x 102 mm) wood stud wall, then the wall would have a one-hour fire resistance and could be used in a load-bearing capacity (Figure 9.14).

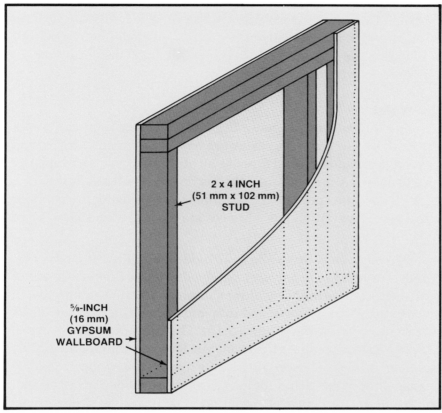

**2 x 4 INCH
(51 mm x 102 mm)
STUD**

**⅝-INCH
(16 mm)
GYPSUM
WALLBOARD**

Figure 9.14 A wall covered with 5/8-inch (16 mm) gypsum wallboard has a fire resistance rating of one hour.

Building codes typically require fire-resistive partitions in locations such as corridor walls (one hour), stairway and elevator shaft enclosures (two hours), and occupancy separations (one to four hours).

When the floor area of a building is subdivided with numerous fire-resistive partitions and/or walls, it is said to be compartmentalized. The fire-resistive partitions are designed to contain fire and block its spread. Of course, any openings in the partition, such as doors, windows, and access panels, nullify the value of the partitions unless these openings are protected by fire doors, shutters, or other means. Fire-resistive compartmentalization, while beneficial, provides only passive fire protection. That is, it may block or retard a fire but it cannot extinguish the fire. Either hose streams or automatic sprinklers are necessary to extinguish the fire.

Fire-resistive partitions can be constructed from a wide variety of materials, including wire lath and plaster, gypsum wallboard, concrete block, and combinations of materials. The degree of fire resistance provided depends on the material used and its thickness. Thus, in the previous example, if two layers of ⅝-inch (16 mm) gypsum wallboard were applied to both sides of the steel studs, the fire resistance would be increased to two hours.

It should be noted that gypsum wallboard partitions, while capable of providing good fire resistance, are relatively easy to penetrate with forcible entry tools. This knowledge can be helpful in reaching the seat of a fire in a compartmented building, such as a modern apartment building.

Fire Walls

The difference between a fire-resistive partition wall and a fire wall is a matter of degree. A fire wall is designed to withstand a severe fire exposure and to act as an absolute barrier against the spread of fire. A fire wall usually has fire resistance of four hours, with all openings protected by automatically closing fire doors. A fire-resistive partition has a fire resistance less than that of a fire wall.

The purpose of fire walls is to divide a building into compartments so that a fire in one area will be limited to that area and not destroy the entire building. For example, a 100,000-square foot (9 290 m²) factory can be divided into four 25,000-square foot (2 323 m²) areas by fire walls (Figure 9.15). The containment of a fire to one area greatly reduces the economic loss and enables a stricken business to recover more quickly. Fire walls can also separate various functions within a plant so that loss of one function will not result in total loss of a facility. For example, a hazardous

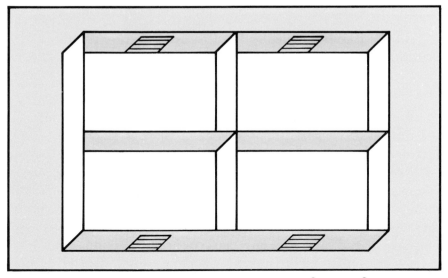

Figure 9.15 Fire walls will help to contain fire to one 25,000 ft.² (2 323 m²) section of this 100,000 ft² (9 290 m²) building.

chemical blending operation can be separated from shipping, warehousing, and other departments.

Building codes usually limit the maximum area of various types of buildings as a means of protecting a community from a conflagration. A typical municipal code restricts two-story mercantile buildings of wood-joisted construction to 8,000 square feet (743 m^2) unless the buildings are equipped with an automatic sprinkler system. If a larger building is desired, fire walls must be provided to limit the maximum individual areas to 8,000 square feet (743 m^2).

Again, the primary concern when penetrating a fire wall is that a breach in the wall may allow fire to travel from one compartment to the next. Hoselines should always be placed in the interior to prevent this possibility.

Fire walls are always constructed of masonry materials. They are designed to be self-supporting so that structural collapse on either side of the fire wall will not damage it. Fire walls are constructed of brick 8 inches (203 mm) thick, solid concrete block 8 inches (203 mm) thick, hollow tile 10 inches (254 mm) thick, or combinations of these materials. Because fire walls are intended to be absolute barriers against fire, no combustible construction is permitted to penetrate the wall. Combustible floor and roof beams that abut a fire wall cannot pass through the wall. In addition, the fire wall must extend vertically above combustible walls and roofs to prevent the radiant heat of flames from igniting adjacent surfaces. This protection is accomplished by topping the fire wall with a parapet (Figure 9.16). The parapet height above the combustible roof is determined by the building code. A code may require parapets 18 to 36 inches (457 mm to 914 mm) in height.

Figure 9.16 A parapet is an extension of a fire wall above the roof that is designed to prevent the radiant heat from fire from igniting adjacent surfaces.

A properly constructed and maintained fire wall is a powerful ally to tactical fire fighting forces. When a section of building on one side of a fire wall becomes heavily involved, the fire wall is a natural line along which to establish a defense. With fire doors closed, one or two handlines can be positioned to check for any spread of fire at cracks or around door edges. This can be accomplished with a minimum of personnel, freeing other firefighters to protect exposures or attack the main body of fire. Great care must be exercised in opening fire doors during the course of fighting fire. Should the situation become untenable, forcing firefighters to withdraw, fire doors must be closed.

A breached opening through a fire wall can be used as a point from which to attack the fire. A primary consideration when deciding whether to breach a fire wall, however, is that the opening cannot be reclosed, as with a fire door. Such action is irreversible in the event that conditions deteriorate and firefighters must retreat from their position. Open and unprotected, the breached wall would then allow fire to travel from the area of fire involvement into the uninvolved area on the opposite side of the wall.

OPENING WALLS

The decision to open a wall should be based on a number of factors. These include, but are not limited to, the following:

- Tactical requirements
- Door and window accessibility
- Damage potential
- Wall materials
- Load-bearing function
- Fire-resistive function
- Stability

Once the decision has been made that through-the-wall entry should be performed, tools are chosen to carry out the method of entry. The entry method, of course, is based on the type of wall construction.

Opening Masonry Walls

Opening, or breaching, a masonry wall can be done with several tools. One of the most popular breaching devices is the battering ram. Made of iron or steel and equipped with handles and hand guards, the battering ram is designed to be operated by several people. Some rams can be used from either end. One end is pointed to break concrete, brick, and stone, while the opposite end is rounded for battering doors. Two people can use a battering ram to handle such barriers as brick walls, but concrete or block walls may require four people using a larger ram.

As with any forcing operation, full protective clothing should be worn during battering operations. Eye protection is particularly important because chips of masonry tend to fly from the point of impact. The battering ram should be heavy enough to accomplish the job without requiring that it be swung with exceptional force.

If a masonry wall is breached to gain access to a door lock, the hole will be made immediately adjacent to the door frame near the lock (Figure 9.17). This opening allows personnel to reach inside to unlock the mechanism. This is a preferred method of breaching because damage is minimized.

If a larger opening in the wall is required, as when opening a wall to allow passage of personnel and equipment to the interior, extra precautions must be taken. Be sure that a hole of the required size will not weaken the wall to the point that it becomes unstable. This is especially important if it is a load-bearing wall. Once the function of the wall has been determined, the operation can begin. Make the hole large enough to allow passage of personnel equipped with self-contained breathing apparatus and tools. It is suggested that the hole be diamond shaped (Figure 9.18). This means that the opening should be wider in the center than at the top and bottom to permit passage of the widest part of the body. Remember that masonry walls usually contain steel reinforcement ("re-bar"). This requires that the steel be cut, either with a circular saw or a cutting torch.

Firefighters can also use power tools such as air chisels and hydraulic spreaders for breaching masonry and concrete walls. Once the initial opening has been made, continue to widen the hole until it is large enough for firefighters to enter. If a brick wall has a plaster stucco finish, remove the plaster stucco before breaching the wall.

Figure 9.17 A small breached hole near the door frame permits reaching inside to unlock the door.

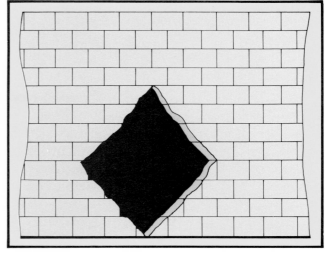

Figure 9.18 When breaching a wall, make the hole diamond shaped and large enough to permit passage of fully equipped personnel.

Opening Metal Walls

Before cutting a metal wall, examine the panels to determine the location of studs or other supports. Also determine if the wall is load bearing. As with any hollow wall, anticipate that there may be electrical wiring and plumbing beneath the exterior surface. This requires that initial cuts be made carefully to avoid contact with such obstacles.

One of the best tools for opening a metal wall is a power saw with an aluminum oxide metal-cutting blade. The size and location of the cut will vary, depending on the purpose of the opening. If only a small access hole will enable firefighters to reach in and unlock a door, a triangular cut should suffice. Make the right and left sides of the cut, then fold the metal down at the bottom. Larger holes may require cutting along metal studs to create an opening large enough to allow passage of fully equipped personnel. If a stud must be cut to make a large enough opening, make sure that wall stability is not compromised. **CAUTION:** Be careful when opening metal walls covered with porcelain. The porcelain will chip and fly off as the cut proceeds through the underlying metal. Be sure to wear eye protection and keep unnecessary personnel out of the cutting area.

After cutting the exterior metal panel, remove the metal and place it in a safe location so that sharp edges will not endanger personnel. Remove any insulation or other obstructing material from the wall opening and place it in the same location. Then examine the inside wall surface to determine the best way to remove it. In many cases, an axe can be used to cut through the interior covering if it is gypsum wallboard or another such nonmetal material. If the covering is thin-gage metal, an axe can be used, but a power saw is still the best tool.

The cutting torch is another tool that is effective for opening metal walls. It should be noted, however, that cutting with this tool may ignite insulating materials within the wall. The danger of cutting electrical wiring, conduit, and plumbing is always present, so due caution should be taken.

Opening Wood-Frame Walls

Wood-frame walls can be relatively easy to open because they are usually made of materials that can be pulled or cut free. Difficulties during forcible entry come from the fact that the walls contain several layers of materials, as well as pipes and electrical wiring. Not only do pipes and wiring present a hazard when walls are being opened, they also block easy entrance once an opening is made.

When cutting a wood-frame wall, as with any other wall, determine whether it is load bearing. If it is, the size of the opening should be restricted. It is not a good idea to cut studs in a wall that supports the weight of heavy interior loads. There should be no problem, however, in cutting a single stud to make a hole large enough to permit passage of personnel. This should be regarded as the limit in breaching a load-bearing wall unless shoring is provided.

Use prying tools or axes to remove such coverings as shingles or vertical siding. Cut long, horizontal siding with a power saw so that an easily repairable hole is created. Making this type of opening requires careful attention to prevent accidental cutting of studs, piping, and wiring. Once the exterior siding is removed, sheathing and other insulating materials must also be cut and pulled free to reveal the stud framework of the wall. Interior wall coverings such as gypsum wallboard can be cut or punched through with axes or other tools. Pipes and wiring should be left in place. If it is necessary to cut and remove a stud, make the initial cut slightly overhead, then push or pull the lower stud section free of the bottom plate if siding has been removed down to that level. If not, cut the stud again at the bottom of the opening.

In commercial buildings that have been remodeled, it is common to find several layers of siding or even new walls on the exterior of wood-frame walls. This process of building over existing walls is called "wrapping." New walls may even be constructed of metal or masonry, requiring the use of a number of different tools to penetrate the double wall layer. The best way to deal with this problem is to be prepared by pre-incident familiarization of the structure.

Opening Partition Walls

Partition walls are usually much easier to open than exterior walls because they do not have to be weatherproofed with such materials as insulation and siding. Hollow walls are covered with such materials as gypsum wallboard, lath and plaster, or wood paneling. The hazard of electrical wiring and piping still exists, but the walls usually yield to previously described entry techniques.

The important thing to remember when opening a partition wall is that the wall may be a barrier to fire extension from the opposite side. Breaching the wall allows fire to move from one side of the partition to the other, so due caution should be taken (hoselines in place; follow-up protection of the opening).

Before opening the wall, check for heat around the area chosen for the opening. Partition walls are usually good conductors of heat because they lack insulation. A check for heat can be made by observing discoloration of paint and wallpaper or by feeling the wall. If the wall is unusually warm, it may indicate that fire is either on the other side or within the wall itself. Hand-held electronic infrared scanning devices can also be used for detecting hidden fire (Figure 9.19).

If heat is detected in the area, proceed with caution if the opening is still deemed necessary. Make openings only as large as necessary and protect interior furniture and equipment with salvage covers or by moving them out of the area.

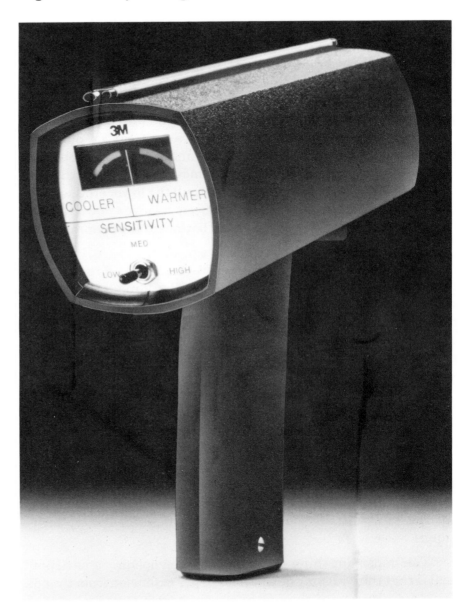

Figure 9.19 A hand-held infrared heat scanner is a valuable aid to detecting hidden fire within walls. *Courtesy of 3M Company.*

Complete the following statements with words or phrases that will make the statement correct:

1. A building is structurally supported by either its _____ or by a/an _____.

2. A load-bearing wall is a wall used for _____

3. An interior non-load-bearing wall is called a/an _____ wall; an exterior non-load-bearing wall is called a/an _____ wall.

4. Exterior masonry walls have a fire resistance rating of _____ to _____.

5. When the floor area of a building is subdivided with numerous fire-resistive walls, it is said to be _____

Answer each of the following questions in a few words or short phrases:

6. How do the following walls have a negative effect on the fundamental fire safety of a building?

   A. Walls with voids: _____

   B. Interior combustible walls: _____

   C. Curtain walls: _____

7. Which wood-frame structure would be more fire resistive, balloon frame or platform frame? Why?

8. In which building would the rate of fire acceleration be greater, a well-insulated building or an uninsulated building? Why?

Determine whether the following statements are true or false. If false, state why:

9. A primary concern when considering through-the-wall entry to a building is whether structural integrity will be affected.

   ☐ T   ☐ F _____

Review
Answers on page 265

10. Only exterior walls can be load-bearing walls.

☐ T   ☐ F   _____

_____

11. Regarding precast concrete walls:
    A. Precast concrete walls are always load bearing.

☐ T   ☐ F   _____

_____

    B. Solid panels are fire resistive, while sandwich panels are not.

☐ T   ☐ F   _____

_____

    C. A single panel can weigh 20 tons.

☐ T   ☐ F   _____

_____

12. It is difficult to determine visually whether the brick walls of a building are load bearing or they are veneer.

☐ T   ☐ F   _____

_____

13. When fire enters a wall space filled with insulation, it will not pass any further because of the retardant properties of the insulation.

☐ T   ☐ F   _____

_____

14. Gypsum wallboard partitions provide good fire resistance but are relatively easy to penetrate with forcible entry tools.

☐ T   ☐ F   _____

_____

15. Fire walls . . .
    A. are rated at a fire resistance of two hours.

☐ T   ☐ F   _____

_____

    B. should have all openings protected by automatically closing fire doors.

☐ T   ☐ F   _____

_____

    C. are wood-framed walls.

☐ T   ☐ F   _____

_____

D. are designed to be self-supporting.

☐ T   ☐ F _____

_____

E. terminate at the under side of a building's roof.

☐ T   ☐ F _____

_____

16. When opening a brick veneer exterior, assume that there is electrical wiring and plumbing beneath the exterior surface.

☐ T   ☐ F _____

_____

17. Before opening a partition wall, check for heat around the area chosen for opening.

☐ T   ☐ F _____

_____

Select the choice that best completes the sentence or answers the question:

18. Whether a new building will be structurally supported by the walls or by a structural frame is usually determined by the
   A. height of the structure
   B. maximum floor area of the structure
   C. type of material used for construction
   D. fire codes in effect for that locale

19. If interior walls serve only as partitions, interior structural loads are supported by
   A. columns
   B. floors
   C. exterior walls
   D. none of the above — interior walls are always load bearing

20. Building codes typically require fire-resistive partitions in locations such as
   A. corridor walls   C. occupancy separations
   B. stairway enclosures   D. all of the above

21. After considering all factors when making a decision to do through-the-wall entry, the method chosen should be based on the
   A. amount of time available
   B. number of personnel available
   C. type of wall construction
   D. fire resistance rating of the wall

22. A battering ram is best used to open
    A. a masonry wall
    B. a metal wall
    C. a wood-frame wall
    D. all of the above

23. If it is necessary to make a hole in a masonry wall large enough to allow the passage of fully equipped personnel, it should be in the shape of a/an
    A. diamond
    B. rectangle
    C. triangle
    D. inverted triangle

24. Shoring should be provided whenever more than _____ is/are cut in a load-bearing wall.
    A. one stud
    B. two studs
    C. three studs
    D. none of the above — load-bearing walls should not be breached

Special
Situations

10

NFPA STANDARD 1001
FORCIBLE ENTRY
Fire Fighter I

3-2 Forcible Entry
3-2.1 The fire fighter shall identify and demonstrate the use of each type of manual forcible entry tool.

Fire Fighter II

4-2 Forcible Entry
4-2.1 The fire fighter shall identify materials and construction features of doors, windows, roofs, floors, and vertical barriers and shall define the dangers associated with each in an emergency situation.

4-2.2 The fire fighter shall identify the method and technique of forcible entry through any door, window, ceiling, roof, floor, or vertical barrier.

Chapter 10
Special Situations

Property owners and occupants faced with a high risk of break-in will usually take measures beyond protecting their buildings with well-built and heavily locked doors and windows. Some of these measures, however, present special problems to firefighters. This chapter will describe several of these problems and suggest means to overcome them in the most expedient way.

FENCES

Fences are the most common property barrier and vary widely in design and materials as well as heights. They present a problem that must be overcome rapidly and efficiently because they not only prevent firefighters from reaching structures that lay beyond, but also fire fighting apparatus and equipment.

Applying the same reasoning when sizing up a locked structure, the most logical method to go through a fence is to force the gate, if locked. If this fails, then it will be necessary to either climb over the top of the fence or create an opening large enough to allow the entrance of personnel and equipment.

Depending on the height and design of the fence, climbing over the top can be relatively easy or difficult. The primary concern when moving over a fence is that injury does not occur because of a slip or fall. For this reason, it is recommended that a ladder of appropriate length be used for transversing the fence. This is especially important when considering that a number of personnel loaded with protective clothing, SCBA, and equipment may follow the same route. Climbing with a heavy load can be a difficult task, so a well-placed ladder can make the task much easier.

One of the best ladders for fence work is a small, multipurpose (combination) ladder that either extends or folds out as an A-frame or step ladder. In the A-frame configuration, some models of this ladder can bridge fences of up to six feet (1.8 m) in height (Figure 10.1). Another method is to place attic ladders on each side of the fence so that personnel can step from one to the other at the top (Figure 10.2). Be sure all ladders are well secured to the fence to prevent slipping. Higher fences require longer ladders. Roof ladders of appropriate length, with hooks extended, are ideal for traversing high security fences. As with any ladder work, spot ladders for the best climbing angle, which is 65 to 75 degrees.

Figure 10.1 In the A-frame configuration, a multipurpose (combination) ladder bridges fences up to 6 feet (1.8 m) tall.

Figure 10.2 Secure attic or combination ladders to each side of the fence so that personnel can step from one to the other at the top.

Techniques for opening fences, in many cases, are similar to those for opening walls. Wooden fences can often be simply pried apart to create the necessary size of opening (Figure 10.3). The advantage in this method is that it can be done with virtually no damage and can be easily repaired before leaving the scene. Cutting is another method that can be accomplished quickly and efficiently with power equipment such as circular saws or chain saws. Make all cuts to the wooden covering, but leave the underlying support members intact, if possible, to reduce repair costs.

Fences or walls of masonry material can be breached with battering rams, sledges, or power equipment such as air chisels and jack hammers. This does extensive damage to the barrier and should be done only if gate or over-the-top entry is not practical.

Wire mesh fences (often called "cyclone fences") can be cut with bolt cutters or wire cutters (Figure 10.4). Make all cuts vertically from bottom to top, and avoid cutting directly against a post (repairs are more easily done later if enough fence is available on both sides of the cut to apply a fence stretcher). Have a second person present, if possible, to prevent the fence sections from suddenly recoiling apart as the last cut is completed. These fences are stretched under tension and can spring back dangerously when cut apart.

Figure 10.3 Wooden fences can often be pried apart to create an opening of sufficient size to permit passage of hoselines and equipment. *Courtesy of Loren Dunlap.*

Figure 10.4 Cut a wire mesh fence vertically from bottom to top with bolt cutters or wire cutters. A second person should hold the sections to prevent them from suddenly recoiling as the last cut is completed. *Courtesy of Mary McKinley.*

GATES

As stated before, gates present the best entry point for property barriers such as fences. Gates operate in much the same manner as doors, usually by swinging or sliding. Many are operated manually, but large, heavy gates such as found across driveways are often electrically operated.

Forcing a gate will often involve overcoming some sort of locking device. Simple gates secured with padlocks will require cutting, spreading, or striking the padlock or its chain, if so mounted (see Chapter 6, Forcing Locks). Once the lock is removed the gate can be opened in a normal manner.

Some gates open automatically when an opening device is activated. Examples of such devices are plastic card readers, coded push-button activators, radio or remote-control openers, and key-activated systems. These devices work either to release the locking device or to electrically open the gate. In some installations, both functions occur simultaneously. Pre-incident planning should include identification of the activating device and, if possible, procurement of the means to open the gate (for example, the card, key, or push-button combination).

Electric gates open and close with a motor that drives some type of an operating mechanism, both of which are always located on the interior of the barrier. Swinging gates usually operate with a connecting rod or arm that moves between the motor and the gate (Figure 10.5). Sliding gates move on rollers and may use one of several means to transmit power from the motor to the gate. One of the most common systems utilizes a sprocket-and-chain arrangement that rolls open and closes the gate on a track (Figure 10.6). Another system uses a roller on the motor to contact a rail on the gate, moving the gate as it turns in either direction. Most electric gates have locking mechanisms built into the motor that prevents the gate from being forced open.

Figure 10.5 Swinging gates operate with a connecting rod or arm that moves between the motor and the gate. *Courtesy of C. Eastwood; California Department of Forestry, Pebble Beach, CA.*

Figure 10.6 This sliding gate uses a sprocket-and-chain arrangement that moves the gate on a track. *Courtesy of California Department of Forestry, Pebble Beach, CA.*

Before forcing a gate, check around the motor cover for an external override switch (Figure 10.7 on next page). Some models contain such a switch so that the gate can be operated from the interior of the property without the normal activating device, such as a card or remote switch. Other models may allow a manual crank to operate the gate. This allows the gate opening mechanism to be cranked open or closed in the event of a power failure. If a crank handle is found, insert it into the appropriate opening in the motor case and rotate it to open the gate (Figure 10.8 on next page).

To determine the best way to force an electric gate, the operating mechanism must be identified and action taken on one of its components. In the case of a swinging gate, first check for a locking device on the gate edge where it meets the post. If there is a locking mechanism, prying, cutting, or through-the-lock methods should be used to force it open. Then, it is sometimes only necessary to pull back on the arm, which overcomes the motor

Figure 10.7 Before forcing a gate, check around the motor cover for an external override switch. Activate the switch to open the gate. *Courtesy of Bob Orchard, Kern County (CA) Fire Dept.*

Figure 10.8 Insert the crank handle into the appropriate opening in the motor case. Rotate it to open the gate. *Courtesy of California Department of Forestry, Pebble Beach, CA.*

clutch, to swing the gate open (Figure 10.9). If this does not work, removal of a bolt or pin from one section of the swing arm will free the gate to swing open. Some models contain a frangible disk at the articulating joint (if a two-piece arm) that will break loose if struck sharply with a striking tool (Figure 10.10). If none of these methods work or are possible, the arm can be cut with a circular saw equipped with a metal cutting blade (Figure 10.11). In some cases, one section of the arm is constructed of simple galvanized pipe, which can be cheaply replaced. Make the cut midway on the section as someone holds the gate in place. This will prevent it from pushing against the arm, causing the saw to bind.

Figure 10.9 In some cases, pulling the swing arm is all that is necessary to swing the gate open. *Courtesy of Kern County (CA) Fire Dept.*

Figure 10.10 Driving a bolt or pin from one section of the swing arm frees the gate to swing open. *Courtesy of Kern County (CA) Fire Dept.*

Figure 10.11 Cut the swing arm with a circular saw equipped with a metal-cutting blade. *Courtesy of Kern County (CA) Fire Dept.*

Chain-driven sliding gates can be forced in several ways. One method is to lift the chain, if loose enough, from the sprocket with a prying tool. Once free of the sprocket, the gate should roll open freely. Another method is to cut the chain with a heavy set of bolt cutters, which allows the gate to roll freely. The cut must be positioned at a place on the chain that releases tension to the gate. If the cut is made on the wrong side of the motor, the gate will not release and a second cut will be required.

BARBED WIRE AND RAZOR RIBBON

High-security areas will often be surrounded with fences equipped with strands or coils of barbed wire to prevent intrusion from fence climbers. Razor ribbon wire, a fence-top barrier even more formidable than barbed wire, was originally developd for military and prison security applications. It is now sold and distributed to the private sector and may be found in many high-crime and high-security areas. In some areas, razor ribbon wire is also called "concertina wire."

Razor ribbon presents a serious hazard to firefighters, and for this reason should be avoided if possible. It was designed to be placed in loose coils on either fence tops or the ground around the perimeter of the protected property (Figure 10.12). Anyone attempting to climb over or through the coils will almost always become entangled by the razor-sharp barbs that protrude from the ribbon wire. The razors easily penetrate clothing and cut deeply into the skin. The more a person struggles to remove the barbs, the more the situation worsens. It is almost impossible for anyone to escape the entanglement without assistance. Several cases have been documented in which intruders have become so entangled and severely cut that they bled to death before discovery.

Figure 10.12 Razor ribbon is usually placed in loose coils on fence tops or on the ground around the perimeter of the protected property. The razors easily penetrate clothing and cut deeply into the skin, making it almost impossible for anyone to escape the entanglement without assistance. *Courtesy of American Security Fence Corp.*

One of the most recent problems with razor ribbon is that property owners are placing it in and around buildings in a manner for which it was not intended. Fire escapes in deserted tenement buildings, for instance, have been covered with razor ribbon to prevent intrusion by vagrants (Figure 10.13). This makes access to firefighters very difficult, if not impossible. The problem is compounded when a fire occurs in the structure and vagrants who have managed to penetrate the structure are trapped within.

Another unique application for razor ribbon in high-crime areas is to place it in ductwork and above ceilings in an attempt to prevent intrusion. Needless to say, firefighters must be aware of the potential danger, especially when considering the potential for ceiling collapse by the combined weight of the ribbon and water from above.

Rooftops may also be covered with razor ribbon, especially around vents and skylights, as well as along parapet walls (Figure 10.14). This presents a serious hazard, especially at night when visibility is poor and firefighters must step from ladders to roofs for ventilation operations.

WARNING
When making access to a roof that has razor ribbon wire around the perimeter, always establish at least two escape routes.

Figure 10.13 Fire escapes in deserted tenement buildings are sometimes covered with razor ribbon to prevent intrusion by vagrants. *Courtesy of Bob Ramirez, Los Angeles City Fire Dept.*

Figure 10.14 Rooftops may have razor ribbon along parapet walls. *Courtesy of Bob Ramirez, Los Angeles City Fire Dept.*

As stated before, the best approach with razor ribbon is to avoid it. If it must be penetrated, at least two people would work as a team to cut the materials with metal shears or bolt cutters.

Be aware that the wire, although loosely coiled, is under some tension and may recoil when cut free. Some installations also include a center strand of heavy wire under tension that requires cutting with bolt cutters. For both types of wire, someone should stabilize the wire on each side of the cut so that the sections can be carefully moved back after the cut is completed (Figure 10.15).

Barbed wire should be handled just as carefully as razor ribbon, especially if mounted in coils. It can be cut with wire cutters or bolt cutters, and firefighters should work in teams to accomplish the task safely. In all cases, full protective clothing, including heavy gloves and eye protection, should be worn when cutting razor ribbon and barbed wire. If any cuts are incurred during the operation, immediate medical attention, including tetanus shots, should be sought.

Figure 10.15 Two people should work as a team to cut razor ribbon wire with metal shears or bolt cutters. Because the wire is under some tension and may recoil when cut free, stabilize the wire on each side of the cut with tools or with gloved hands.

Another hazard with razor ribbon and barbed wire is that it may be electrified. This may be done purposely, as an added security measure, or may occur because of downed electrical lines during an emergency incident (Figure 10.16 on next page). In either case, metal fences and security wires should be approached with extreme caution. According to law, electrified fences must be prominently posted with appropriate warning signs, but firefighters should not depend on the presence of this type of warning.

Figure 10.16 Razor ribbon and barbed wire may become electrified when overhead electrical lines fall across the fence during an emergency incident. In this photograph, a section of razor ribbon wire has dropped across power lines, electrifying the entire fence. *Courtesy of Bob Ramirez, Los Angeles City Fire Dept.*

WARNING
Do not attempt to cut a fence or wire without confirming that it is electrically safe.

The best way to identify charged fences is **before** an incident, through contact with property owners.

GUARD DOGS

Another barrier to intruders that presents a real hazard to firefighters is the guard dog. Business owners often contract with licensed guard dog contractors who provide the animals for a set fee. The dogs are brought in shortly after the business closes and are released within a fenced area or within the building. The dogs roam freely and are trained to be extremely attentive to sounds and movements within the property. Some breeds will give no warning before attacking an intruder, and are capable of causing serious physical wounds to anyone encountered (Figure 10.17).

Some localities require signs to be posted that warn of the presence of guard dogs (Figure 10.18), but this is not always done. Although most dogs make their presence known as soon as forcible entry crews start operations on doors, windows, or fenced enclosures, this may not always be the case. The best way to prepare for an encounter with an aggressive animal is to be aware of its presence before approaching the premises. This can only happen through pre-incident interviews with business owners, who should be questioned about such security measures. Some owners are hesitant to volunteer such information, so a direct question must be used to ascertain whether dogs are used.

Figure 10.17 Some guard dogs give no warning before attacking. (This picture was staged with a trained guard dog; the firefighter is wearing an arm guard for protection.) *Courtesy of George Braun, Gainsville (FL) Fire Department.*

Figure 10.18 Warning signs are *sometimes* present on gates or doors of property protected by guard dogs.

There is no best way to deal with a guard dog if one is encountered after forcing entry to a property. It is always advisable to have a tool in hand when entering a building, and this can be used for protection. Perhaps one of the most effective tools is a charged hoseline. Few animals can get past a well-aimed straight stream of water and, although an effective defensive weapon, it will not harm the dog.

NONDESTRUCTIVE ENTRY WITH KEY BOXES

No training manual about forcible entry is complete without some mention of a method that gives firefighters a better option than forcing doors, windows, or locks to enter a building. This method, quite simply, is to unlock the building with the same keys the occupant uses.

The greatest problem with this method is that it requires carrying large numbers of keys in an apparatus (Figure 10.19). This becomes impractical because it is nearly impossible to find and use a single key from such a collection that could number in the hundreds. Another problem is that occupants change locks periodically, which requires issuing new keys. It would be difficult to ensure that all keys are current.

These problems are solved when a simple device called the key box is mounted on the outside of a building. A key box is designed to hold all the keys that open external and internal doors of the building. Most types are of heavy-duty construction and can be either mounted on a wall (surface-mounted) or inset (flush-mounted) (Figure 10.20 on next page). To prevent easy access by intruders, the boxes are often mounted high on the wall so that a

Figure 10.19 Carrying large numbers of keys in an apparatus is impractical.

Figure 10.20 Key boxes can be either surface-mounted (left) or flush-mounted (right) on the wall. *Courtesy of The Knox Company, Newport Beach, CA.*

ladder, such as carried by the fire department, is required to reach them (Figure 10.21). Boxes can also be equipped with alarm systems to alert police or security personnel if entry is made to the box.

Figure 10.21 To prevent easy access by intruders, key boxes are often mounted high on the wall so that a ladder is required to reach them. *Courtesy of The Knox Company, Newport Beach, CA.*

The key collection within the box is maintained by the occupant or owner so that the keys are always current. Key boxes can be opened with a master key, which is held only by the fire and/or police department, and which opens all key boxes of that type in the area. This enables fire personnel to access keys to any building in their jurisdiction with only one key. Occupant keys fit only the key box for the building that person controls, so that no person owning or maintaining a key box can gain access to other boxes. The master key is the only key that can open all key boxes.

Needless to say, the key box has many advantages. In addition to eliminating the necessity of carrying many keys in fire apparatus, it reduces entry time into a locked building. It prevents needless damage to doors, windows, and locks because any external or internal door can be unlocked by fire personnel as needed. This is especially advantageous in large buildings such as high-rise offices, where many separate sections of the building may need to be unlocked and searched.

After an incident, the building can be relocked and keys returned to the key box. An inventory list is usually maintained within the box so that if several keys were removed from the key box an accounting can be made to ensure all keys are returned.

Review

Answers on page 265

Complete the following statements with words or phrases that will make the statement correct:

1. The two methods by which gates operate are by _____ or by _____.

2. It has been decided to force open a locked electric gate. You have not been able to unlock the gate but you have reached the motor that controls the gate. Before forcing the gate, you should check to see if there is a/an _____ or a/an _____.

Determine whether the following statements are true or false. If false, state why:

3. Through-the-lock entry is a recommended method of forcing a manually operated gate.

   ☐ T   ☐ F   _____
   _____

4. The locking mechanism on a swinging electric gate has been forced. The best method of forcing the gate open is to cut the connecting arm between the motor and the gate.

   ☐ T   ☐ F   _____
   _____

5. One method of forcing a chain-driven sliding gate is by cutting the chain.

   ☐ T   ☐ F   _____
   _____

6. When using a roof ladder to climb a high security fence, the hooks should be extended, and the ladder should be hung vertically from the top of the fence.

   ☐ T   ☐ F   _____
   _____

7. The recommended techniques for opening fences are similar to those for opening walls.

   ☐ T   ☐ F   _____
   _____

8. When cutting barbed wire, firefighters should work in teams and wear full protective clothing.

   ☐ T   ☐ F   _____
   _____

9. A water stream can be used as a defense against a guard dog that is discovered after forcing entry to a property.

   ☐ T   ☐ F   _____

   _____

Select the choice that best completes the sentence or answers the question:

10. The best method to access a burning building or property surrounded by a fence with a locked gate is to
    A. climb the fence
    B. force open the gate
    C. cut an opening in the fence
    D. bridge the fence with a step ladder

11. Electric gates may be activated by
    A. plastic card readers
    B. coded push-button activators
    C. remote-control openers
    D. any of the above

12. When cutting a wire mesh or cyclone fence, cut the wire _____ from bottom to top _____ a post.
    A. vertically; adjacent to
    B. diagonally; adjacent to
    C. vertically; away from
    D. diagonally; away from

13. Which type of ladder should be used to climb a fence?
    A. A-frame
    B. Attic ladder
    C. Roof ladder
    D. Any of the above

14. The recommended procedure for dealing with razor ribbon wire on the ground inside a fence is to
    A. avoid it whenever possible
    B. cut it with bolt cutters
    C. lay salvage covers over it
    D. span it with an A-frame ladder

15. A key box that holds the keys to a building
    A. is usually placed just inside the main entry door
    B. can be opened with a master key carried by the fire department or police department
    C. contains the keys to interior, but not exterior, doors
    D. all of the above

Review Answers

CHAPTER 1

1. quickly; with minimum damage
2. A. Ease of access
   B. Vulnerability of assemblies to forcible entry
   C. Minimum potential for damage
   D. Tactical feasibility for rescue and fire attack
3. Interior fire attack and rescue operations cannot be started until forcible entry is completed.
4. True
5. False; skill is developed by actually using each tool
6. False; Well-trained crews can gain entry with minimum damage
7. True
8. True
9. D
10. B

CHAPTER 2

1. A. Prying and spreading tools
   B. Cutting and boring tools
   C. Striking and battering tools
   D. Lock-entry tools
2. A. they develop exceptional force
   B. they work with great speed
3. A. the weight of the tool
   B. the velocity of the impact
4. A. As close as possible
   B. As far away as possible
5. By welding a steel handle to the head
6. Mismatching the tool to the material to be cut
7. Choosing a tool that is too light
8. A. Unsafe
   B. Unsafe
   C. Unsafe
   D. Safe
   E. Safe
   F. Safe
   G. Unsafe
9. A. True
   B. False; Never work under a load only supported by the bags
   C. False; Use insulation to protect the bag
   D. False; Never stack more than two bags
10. A. True
    B. True
    C. True

D. False; Valves should be closed to prevent the escape of residual gas and acetone
11. False; A razor-sharp edge tends to chip
12. True
13. D
14. A
15. A
16. D
17. D
18. C
19. C
20. D
21. C
22. A
23. B
24. B
25. A
26. D
27. A
28. C
29. A
30. D
31. A
32. B
33. B
34. B
35. D
36. C
37. A

CHAPTER 3

1. A. Doors
   B. Windows
   C. Locks
   D. Walls
2. A. By the components of which they are constructed
   B. How they operate
   C. By their construction design
3. A. Wood
   B. Metal
   C. Glass
4. A. SC
   B. B
   C. SC
   D. HC
5. A. Swinging
   B. Sliding
   C. Folding
   D. Rolling
   E. Revolving

6. Free-swinging, because its locking mechanism is less complex
7. It will not allow the passage of hose to the inside
8. A. Horizontal and vertical sliding
   B. Single and double swinging
   C. Overhead rolling
9. True
10. True
11. D
12. A
13. B
14. right-swinging, in-swinging
15. left-swinging, out-swinging

CHAPTER 4

1. A. Fire location within the building
   B. Distance of entry point from pumping apparatus
   C. Type of door assemblies
   D. Accessibility
2. Fire location, because this has the most bearing on tactical feasibility
3. Attempt to open it in a normal manner
4. A. How is this door constructed?
   B. How does it operate?
   C. In what type of frame is it mounted?
   D. What type of hardware does the assembly contain?
   E. What types of locking devices are in use?
5. The door that opens outward. Because it is easier to pry open than an inward-swinging door; often, too, hinge pins may be removed.
6. Directly on the locking device or on the door at latch level, close to the frame
7. A. Rolling
   B. Folding
   C. Slab

8. Breaking out a panel to reach the locking device
9. False; Working more slowly does not necessarily minimize damage
10. True
11. True
12. False; Doors can be pried on either side
13. False; An in-swinging door can be pried, although two tools are sometimes required; in some cases the stop can be removed to allow access to the latch
14. False; They can be pried
15. False; First attempt to disengage the rod; if this fails, pry it open; break glass only as a last resort
16. True
17. True
18. False; The locking mechanism should be disarmed

19. A	23. D
20. B	24. B
21. A	25. D
22. D	26. A

CHAPTER 5

1. A. Bored or cylindrical
   B. Mortise
   C. Rim
   D. Pre-assembled or unit
   E. Exit device
2. Because the locks have relatively short latches
3. A. Latch
   B. Deadbolt
   C. Interlocking deadbolt
4. A. Rim — single swinging doors
   B. Mortise — single swinging doors
   C. Surface vertical rod — double swinging doors
   D. Concealed vertical rod — double swinging doors

5. The higher lock. Because it is an auxiliary lock and is usually the more substantial of the two.
6. A. Rim lock
   B. Tubular deadbolt
7. True
8. False; They are usually cut-resistant
9. True
10. C
11. A
12. D
13. D
14. B
15. B
16. D
17. E
18. C
19. A

CHAPTER 6

1. Retracting the latch on a lock with a thin, flexible object
2. No. Because the deadlatch prevents the deadbolting latch from clearing the strike.
3. Yes
4. Yes, by punching the inside portion of the lock from the door
5. In the center
6. A. Cut the door in half from top to bottom
   B. Cut a triangle from the door and reach in to remove the bar from its bracket
7. Mortise, rim
8. False; Other methods may be less damaging on some doors
9. True
10. False; Many tools are inexpensive; some can be made in the shop
11. True
12. False; These locks are resistant to forcible entry
13. False; Cylinder guard bolts can be sheared with a cutting tool

14. True
15. False; Since the chain is not rigid, the padlock is not easily broken
16. True
17. A. True
    B. True
    C. False; It is not designed to cut padlock components
    D. False; There is no wedge for spreading the shackle
    E. False; There is no drill bit on the tool
    F. True
18. C
19. B
20. C
21. B
22. B
23. D
24. C
25. A
26. B

CHAPTER 7

1. A. Resistance to breakage
   B. Pattern of breakage
   C. Expense of replacement
2. A. Plate glass
   B. Float glass
3. A. The primary function for which they were designed
   B. The manner in which they operate
4. A. wood
   B. aluminum
5. A. Fixed
   B. Sliding
   C. Swinging
   D. Pivoting
   E. Security

6. False; Window entry may be required for tactical reasons
7. False; They are swinging windows
8. True
9. C
10. A
11. B
12. D
13. D
14. C
15. C
16. A
17. D
18. C
19. B

CHAPTER 8

1. A. Identification of the best points of entrance
   B. Possible methods to gain entry
   C. Tools required to carry out the operation
2. A. Type of glazing
   B. Type of latch or lock
   C. Type of opening mechanism
   D. Cost of repair or replacement
3. A. Remove the glazing from the frame
   B. Break the glazing out completely
4. A. The upwind side
   B. The flat side of the axe head
   C. As high as possible
   D. Above
5. Check to see if it is locked
6. A. Loiding
   B. Prying to force the lock
   C. Breaking the glazing to unlock the window

   D. Breaking glazing out of the entire sash
7. A. the side opposite the sash
   B. the side that does not have a hinge
8. True
9. False; it should be cut with a circular saw
10. False; The hands should never be used to break glass
11. True
12. False; They should be removed, not broken
13. A
14. C
15. B
16. B
17. C
18. C
19. D
20. C

CHAPTER 9

1. walls, structural frame
2. structural support
3. partition; curtain or panel
4. two; four
5. compartmentalized
6. A. Provide paths for the spread of fire
   B. Contribute fuel to a fire
   C. Provide little resistance to fire
7. Platform frame. The wall plates serve as fire stops
8. A well-insulated building, because the insulation prevents the escape of heat from the building interior
9. True
10. False; Interior walls can be load bearing
11. A. False; They made be load bearing or non-load bearing
    B. False; Both are fire resistive
    C. True
12. True
13. False; Many kinds of insulation are flammable
14. True
15. A. False; Four hours
    B. True
    C. False; They are constructed of masonry materials
    D. True
    E. False; They must extend above the roof
16. True
17. True
18. A
19. A
20. D
21. C
22. A
23. A
24. A

CHAPTER 10

1. swinging; sliding
2. external override switch; manual operating crank
3. True
4. False; Other, less destructive methods should be attempted first
5. True
6. False; The ladder should be placed against the fence at a normal climbing angle
7. True
8. True
9. True
10. B
11. D
12. C
13. D
14. A
15. B

Index

IFSTA MANUALS

FIRE SERVICE ORIENTATION & INDOCTRINATION
History, traditions, and organization of the fire service; operation of the fire department and responsibilities and duties of firefighters; fire department companies and their functions; glossary of fire service terms.

FIRE SERVICE FIRST RESPONDER
Covers all objectives for U.S. DOT First Responder Training Course as well as NFPA 1001 Emergency Medical Care sections. The special emphasis on maintenance of the ABC's features updated CPR techniques. Also included are scene assessment and safety, patient assessment, shock, bleeding control, spinal injuries, burns, heat and cold emergencies, medical emergencies, poisons, behavioral emergencies, emergency childbirth, short-distance transfer, and emergency vehicles and their equipment.

FIRE SERVICE FIRST AID PRACTICES
Brief explanations of the nervous, skeletal, muscular, abdominal, digestive, and genitourinary systems; injuries and treatment relating to each system; bleeding control and bandaging; artificial respiration, cardiopulmonary resuscitation (CPR), shock, poisoning, and emergencies caused by heat and cold; fractures, sprains, and dislocations; emergency childbirth; short-distance transfer of patients; ambulances; conducting a primary and secondary survey.

ESSENTIALS OF FIRE FIGHTING
This manual was prepared to meet the objectives set forth in levels I and II of NFPA, *Fire Fighter Professional Qualifications, 1981*. Included in the manual are the basics of fire behavior, extinguishers, ropes and knots, self-contained breathing apparatus, ladders, forcible entry, rescue, water supply, fire streams, hose, ventilation, salvage and overhaul, fire cause determination, fire suppression techniques, communications, sprinkler systems, and fire inspection.

SELF-INSTRUCTION FOR ESSENTIALS
Over 260 pages of structured questions and answers for studying the *Essentials of Fire Fighting*. Each unit begins with the NFPA Standard No. 1001 required performance objectives. This self-instruction book will help you to learn many of the important topics and review the basic text.

IFSTA'S 500 COMPETENCIES FOR FIREFIGHTER CERTIFICATION
This manual identifies the competencies that must be achieved for certification as a firefighter for levels I and II. The text also identifies what the instructor needs to give the student, NFPA standards, and has space to record the student's score, local standards, and the instructor's initials.

FIRE SERVICE GROUND LADDER PRACTICES
Various terms applied to ladders; types, construction, maintenance, and testing of fire service ground ladders; detailed information on handling ground ladders and special tasks related to them.

HOSE PRACTICES
This new edition has been updated to reflect the latest information on modern fire hose and couplings, including large diameter hose. Details basic methods for handling hose and coupling construction; care, maintenance, and testing; hose appliances and tools; basic methods of handling hose; supply and attack methods; special hose operations.

SALVAGE AND OVERHAUL PRACTICES
Planning and preparing for salvage operations, care and preparation of equipment, methods of spreading and folding salvage covers, most effective way to handle water runoff, value of proper overhaul and equipment needed, and recognizing and preserving arson evidence.

FORCIBLE ENTRY
This comprehensive manual contains technical information about forcible entry tactics; tools; and door, window, and wall construction. Forcible entry methods are described for door, window, and wall entry. A new section on locks and through-the-lock entry makes this the most up-to-date manual available for forcible entry training.

SELF-CONTAINED BREATHING APPARATUS
Beginning with the history of breathing apparatus and the reasons they are needed, to how to use them, including maintenance and care, the firefighter is taken step by step with the aid of programmed-learning questions and answers throughout to complete knowledge of the subject. The donning, operation, and care of all types of breathing apparatus are covered in depth, as are training in SCBA use, breathing-air purification, and recharging cylinders. There are also special chapters on emergency escape procedures and interior search and rescue.

FIRE VENTILATION PRACTICES
Objectives and advantages of ventilation; requirements for burning, flammable liquid characteristics and products of combustion; phases of burning, backdrafts, and the transmission of heat; construction features to be considered; the ventilation process including evaluating and size-up is discussed at length.

FIRE SERVICE RESCUE PRACTICES
Sections include water and ice rescue, trenching, cave rescue, rigging, search-and-rescue techniques for inside structures and outside, and taking command at an incident. Also included are vehicle extrication and a complete section on rescue tools. The book covers all the information called for by the rescue sections of NFPA 1001 for Fire Fighter I, II, and III, and is profusely illustrated.

HAZARDOUS MATERIALS FOR FIRST RESPONDERS
Designed to assist first-arriving companies in identification of hazardous materials and scene assessment. Covers initial scene control and operations to maintain safety for all responders. Includes characteristics of hazardous materials, identifying hazardous materials, preincident planning, personal protective equipment, command and control of incidents, operations at hazardous materials incidents, and control agents.

THE FIRE DEPARTMENT COMPANY OFFICER
This manual focuses on the basic principles of fire department organization, working relationships, and personnel management. For the firefighter aspiring to become a company officer and the company officer who wishes to improve management skills, this manual will be invaluable. This manual will help individuals develop and improve the necessary traits to effectively manage the fire company.

FIRE CAUSE DETERMINATION
Covers need for determination, finding origin and cause, documenting evidence, interviewing witnesses, courtroom demeanor, and more. Ideal text for company officers, firefighters, inspectors, investigators, insurance and industrial personnel.

PRIVATE FIRE PROTECTION & DETECTION
Automatic sprinkler systems, special extinguishing systems, standpipes, detection and alarm systems. Includes how to test sprinkler systems for the firefighter to meet NFPA 1001.

INDUSTRIAL FIRE PROTECTION
Devastating fires in industrial plants occur at a rate of 145 fires every day. *Industrial Fire Protection* is the single source document designed for training and managing industrial fire brigades.

A must for all industrial sites, large and small, to meet the requirements of the Occupational Safety and Health Administration's (OSHA) regulation 29 CFR part 1910, Subpart L, concerning incipient industrial fire fighting.

CHIEF OFFICER
The role of the fire service has expanded from solely fire suppression to include public education, emergency medical services, and hazardous materials control. This manual provides an overview of the skills needed by today's chief officer. Included are coordination of emergency medical services, master planning, disaster planning, budgeting, information management, labor relations, and the political process. Referenced where appropriate to NFPA 1021, Fire Fighter Professional Qualifications, for levels V and VI.

SI CHIEF OFFICER
This manual assists those reading the text in retaining the principles and concepts necessary to be a skilled chief officer. It includes structured questions and answers, exercises, and a new supplemental section on Resolution by Objectives. It will serve as a review of the text and assist in promotional exam preparation.

FIRE SERVICE INSTRUCTOR
Characteristics of a good instructor; determining training requirements and what to teach; types, principles, and procedures for teaching and learning; training aids and devices; conference leadership.

PUBLIC FIRE EDUCATION
Public fire education planning, target audiences, seasonal fire problems, smoke detectors, working with the media, burn injuries, and resource exchange.

FIRE INSPECTION AND CODE ENFORCEMENT
This revised edition is designed to serve as a reference and training manual for fire department inspection personnel. Includes authority and responsibility; inspection procedures; principles of fire protection and fire cause determination; building construction for fire and life safety; means of egress; extinguishing equipment and fire protection systems; plans review; storage, handling, and use of hazardous materials.

STUDY GUIDE FOR FIRE INSPECTION AND CODE ENFORCEMENT
The **Fire Inspection And Code Enforcement Study Guide** is designed to supplement the **Fire Inspection and Code Enforcement** manual by providing questions and answers to key areas addressed within the manual. This study guide can help the student obtain a thorough understanding of NFPA 1031, levels I and II. Included are case studies that simulate inspection responsibilities.

WATER SUPPLIES FOR FIRE PROTECTION
Designed to help improve understanding of the principles, requirements, and standards used to provide water for fire fighting. Revised (1987) to include information about rural water supplies. Includes water supply management, water system fundamentals, fire hydrants, fire flow testing, static sources, relay operations, and shuttle operations.

FIRE DEPARTMENT PUMPING APPARATUS
The driver/operator's encyclopedia on operating fire department pumps and pumping apparatus. Includes the driver/operator; operating emergency vehicles; types of pumping apparatus; positioning apparatus; fire pump theory; operating fire pumps; apparatus maintenance; apparatus testing; apparatus purchase and specifications. Also included are detailed appendices covering the operation of all major manufacturers' pumps.

FIRE STREAM PRACTICES
Characteristics, requirements, and principles of fire streams; developing, computing, and applying various types of streams to operational situations; formulas for application of hydraulics; actions and reactions created by applying streams under different circumstances.

FIRE PROTECTION ADMINISTRATION
A reprint of the Illinois Department of Commerce and Community Affairs publication. A manual for trustees, municipal officials, and fire chiefs of fire districts and small communities. Subjects covered include officials' duties and responsibilities, organization and management, personnel management and training, budgeting and finance, annexation and disconnection.

BUILDING CONSTRUCTION
This 170-page manual covers building construction features vital to developing fire fighting tactics in a structure. Subjects include construction principles, assemblies and their fire resistance, building services, door and window assemblies, and special types of structures.

FIREFIGHTER SAFETY
Basic concepts and philosophy of accident prevention; essentials of a safety program and training for safety; station house facility safety; hazards en route and at the emergency scene; personal protective equipment; special hazards, including chemicals, electricity, and radioactive materials; inspection safety; health considerations.

AIRCRAFT FIRE PROTECTION AND RESCUE PROCEDURES
Aircraft types, engines, and systems, conventional and specialized fire fighting apparatus, tools, clothing, extinguishing agents, dangerous materials, communications, pre-fire planning, and airfield operations.

GROUND COVER FIRE FIGHTING PRACTICES
Ground cover fire apparatus, equipment, extinguishing agents, and fireground safety; organization and planning for ground cover fire; authority, jurisdiction, and mutual aid, techniques and procedures used for combating ground cover fire.

INCIDENT COMMAND SYSTEM
Developed by a multi-agency task force, this manual is designed to be used by fire, police, and other governmental groups during an emergency. ICS is the approved basic command system as taught at the National Fire Academy. Includes components of ICS, major incident organization, and strike team kind/types and minimum standards.

LEADERSHIP FOR THE COMPANY OFFICER
A 12- to 15-hour course designed for the new company officer or the firefighter who anticipates promotion to company officer. Includes introduction to leadership, leadership techniques, theories of human motivation, determining leadership style, leadership styles, and demanding leadership situations.

FIRE SERVICE PRACTICES FOR VOLUNTEER AND SMALL COMMUNITY FIRE DEPARTMENTS
A general overview of material covered in detail in *Forcible Entry, Ladders, Hose, Salvage and Overhaul, Fire Streams, Apparatus, Ventilation, Rescue, Inspection, Self-Contained Breathing Apparatus,* and *Public Fire Education.*

HAZ MAT LEAK & SPILL GUIDE
A brief, practical treatise that reviews operations at spills and leaks. Sample S.O.P. and command recommendations along with a decontamination guide.

TRANSPARENCIES
Multicolored overhead transparencies to augment *Essentials of Fire Fighting* and other texts. Since costs and availability vary with different chapters, contact IFSTA Headquarters for details.

SLIDES
2-inch by 2-inch slides that can be used in any 35 mm slide projector. Subjects include:

Sprinklers (6 modules)
Smoke Detectors Can Save Your Life
Matches Aren't For Children
Public Relations for the Fire Service
Public Fire Education Specialist (Slide/Tape)

FIREFIGHTER VIDEOTAPE SERIES
Designed to reinforce basic skills and increase knowledge on a variety of fire fighting topics. Excellent for use with *Essentials* or *Volunteer* to review and emphasize different topics. Available for Firefighter levels I, II, and III.

MANUAL HOLDER
The fast, efficient way to organize your IFSTA manuals. These attractive heavy-duty vinyl holders have specially designed side panels that allow easy access to all manuals. Each will hold up to eight manuals.

IFSTA BINDERS
Heavy-duty three-ring binders for organizing and protecting your IFSTA manuals. Available in two sizes: 1½ inch and 3 inch.

WATER FLOW TEST SUMMARY SHEETS
50 summary sheets and instructions for use; logarithmic scale to simplify the process of determining the available water in an area.

PERSONNEL RECORD FOLDERS
Personnel record folders should be used by the training officer for each member of the department. Such data as training, seminars, and college courses can be recorded, along with other valuable information. Letter size or legal size.

ADDITIONAL PUBLICATIONS AND TRAINING MATERIALS ARE AVAILABLE. CALL FOR A FREE CATALOG.

ORDER FORM

ifsta®

2/89

SHIP TO: DATE: _____

NAME _____

CUSTOMER ACCOUNT NO: PHONE _____

STREET ADDRESS (Shipped UPS) _____

CITY   STATE   ZIP _____

SIGNATURE _____

SOCIAL SECURITY NO. OR FEDERAL ID NO. _____

(NOTE: Order cannot be processed without signature.)

☐ VISA   ☐ MASTER CARD

CARD # _____ EXP. DATE _____

Payment Enclosed ☐   Bill Me Later ☐

Send to:
Fire Protection Publications
Oklahoma State University
Stillwater, Oklahoma 74078-0118
800-654-4055
Or Contact Your Local Distributor

Allow 4 to 6 weeks for delivery.

FILL IN THE ITEMS AND QUANTITIES DESIRED

QUANTITY	TITLE	LIST PRICE	TOTAL

Orders **outside** United States, contact Customer Services for shipping and handling charges

Obtain postage and prices from current IFSTA Catalog or they will be inserted by Customer Services.

NOTE: Payment with your order saves you postage and handling charges when ordering from Fire Protection Publications.

SUBTOTAL _____

Discount, if applicable _____
Postage and Handling
if applicable _____

TOTAL _____

TOLL FREE NUMBER
800-654-4055

FAX YOUR ORDER
405-744-8204

Oklahoma call 405-744-5723

COMMENT SHEET

FORCIBLE ENTRY
7th Edition
2ⁿᵈ Printing, 3/89

DATE _____ NAME _____

ADDRESS _____

ORGANIZATION REPRESENTED _____

CHAPTER TITLE _____ NUMBER _____

SECTION/PARAGRAPH/FIGURE _____ PAGE _____

1. Proposal (include proposed wording, or identification of wording to be deleted), OR PROPOSED FIGURE:

2. Statement of Problem and Substantiation for Proposal:

RETURN TO: IFSTA Editor
             Fire Protection Publications
             Oklahoma State University
             Stillwater, OK 74078

SIGNATURE _____

Use this sheet to make any suggestions, recommendations, or comments. We need your input to make the manuals the most up to date as possible. Your help is appreciated. Use additional pages if necessary.